MATHEMATICS AND THE NATURAL SCIENCES
The Physical Singularity of Life

Advances in Computer Science and Engineering: Texts

Editor-in-Chief: Erol Gelenbe *(Imperial College London)*
Advisory Editors: Manfred Broy *(Technische Universität München)*
Gérard Huet *(INRIA)*

Published

Vol. 1 Computer System Performance Modeling in Perspective:
A Tribute to the Work of Professor Kenneth C. Sevcik
edited by Erol Gelenbe (Imperial College London, UK)

Vol. 2 Residue Number Systems: Theory and Implementation
*by Amos Omondi (Yonsei University, South Korea) and
Benjamin Premkumar (Nanyang Technological University, Singapore)*

Vol. 3: Fundamental Concepts in Computer Science
*edited by Erol Gelenbe (Imperial College London, UK) and
Jean-Pierre Kahane (Université de Paris-Sud, France)*

Vol. 4: Analysis and Synthesis of Computer Systems (2nd Edition)
*by Erol Gelenbe (Imperial College London, UK) and
Isi Mitrani (Newcastle University, UK)*

Vol. 5: Neural Nets and Chaotic Carriers (2nd Edition)
by Peter Whittle (University of Cambridge, UK)

Vol. 6: Knowledge Mining Using Intelligent Agents
*edited by Satchidananda Dehuri (Fakir Mohan University, India) and
Sung-Bae Cho (Yonsei University, South Korea)*

Vol. 7: Mathematics and the Natural Sciences: The Physical Singularity of Life
*by Francis Bailly (Physics, CNRS, France) and
Giuseppe Longo (Informatics, CNRS, Ecole Normale Supérieure of
Paris and Ecole Polytechnique, France)*

Advances in Computer Science and Engineering: Texts   Vol. 7

MATHEMATICS AND THE NATURAL SCIENCES
The Physical Singularity of Life

Francis Bailly
Physics, CNRS, France

Giuseppe Longo
*Informatics, CNRS, Ecole Normale Supérieure of Paris
and Ecole Polytechnique, France*

Imperial College Press

Published by

Imperial College Press
57 Shelton Street
Covent Garden
London WC2H 9HE

Distributed by

World Scientific Publishing Co. Pte. Ltd.
5 Toh Tuck Link, Singapore 596224
USA office: 27 Warren Street, Suite 401-402, Hackensack, NJ 07601
UK office: 57 Shelton Street, Covent Garden, London WC2H 9HE

British Library Cataloguing-in-Publication Data
A catalogue record for this book is available from the British Library.

English translation of the Revised version of:
F. Bailly and G. Longo, "Mathématiques et sciences de la nature. La singularité physique du vivant", Hermann, Paris (2006).

Advances in Computer Science and Engineering: Texts — Vol. 7
MATHEMATICS AND THE NATURAL SCIENCES
The Physical Singularity of Life

Copyright © 2011 by Imperial College Press

All rights reserved. This book, or parts thereof, may not be reproduced in any form or by any means, electronic or mechanical, including photocopying, recording or any information storage and retrieval system now known or to be invented, without written permission from the Publisher.

For photocopying of material in this volume, please pay a copying fee through the Copyright Clearance Center, Inc., 222 Rosewood Drive, Danvers, MA 01923, USA. In this case permission to photocopy is not required from the publisher.

ISBN-13 978-1-84816-693-6
ISBN-10 1-84816-693-1

Printed in Singapore.

Preface

This book aims to draw a possibly unified conceptual framework in reference to the current state of the sciences – mainly physics and biology. This framework will be put into close relationship, while not being subordinated, to the analyses of the foundations of mathematics. As a consequence of this framework, we will propose some principles for a modern philosophy of nature and will develop a theoretical approach for certain aspects of biology. This approach, all the while being inspired by physico-mathematical practices and conceptualizations, will clearly distinguish itself from current physical theories in the specification of living phenomena.

Our analyses will involve reflections on the epistemology of mathematics given that questions regarding "truth," "how reason functions," the role of "mathematical theorization/formalization" or "how knowledge is constructed," are highly correlated to questions pertaining to the foundations of this discipline. Referring to history only, this is the lesson we learn as scientists from Plato, Descartes, Kant, Husserl, Wittgenstein, and many other philosophers whose theories of knowledge so often interact with reflections on mathematics. But why would mathematics be one of the pillars of the intelligibility of the world? Why would any philosophy of knowledge or of nature refer to it, while most foundational analyses (Platonism, formalism, logicism, ...) explicitly posit mathematics to be "outside of the physical and biological world"?

According to the analysis which we develop here, mathematics helps, on the one hand, to constitute the very objects and objectivity of the exact sciences, because it is within mathematics that *thought stabilizes itself*. In this way, the foundation of mathematics "mingles" with that of other spheres of knowledge and with their constitutive dynamics: it is the very result of our practices of knowledge. On the other hand, the conceptual stability of mathematics and its relative simplicity (it can be profound all the while stemming from principles which are elementary, sometimes very simple) are at the center of the connection which we will draw with certain elementary cognitive processes. We will mainly refer to those processes which reflect or impose regularities in the world, by organizing it through our active presence within this world as living beings (living also in intersubjectivity and history). It will be thus a question of grasping the role of action in time and space, as well as the organization of these by means of "gestures," as we will put it, and by means of concepts, rich in history and language, both being, since their origin, eminently mathematical concepts. It is for these reasons, in our opinion, that all theories of knowledge have addressed, in one way or another, issues pertaining to the foundations of mathematics, a "purified" knowledge, which is both mysterious and simple, where the analysis of reasoning is performed with extreme clarity and the construction of concepts is rooted upon praxes originating with our humanity.

Symmetrically, a sound epistemology of mathematics must try to make a philosophy of nature explicit. In any case, such a philosophy is implicit to this epistemology because the great choices concerning the foundations of mathematics, logicism, formalism, Platonism, and various types of constructivism, including the "geometrical" perspective that we will adopt, contain an approach to the knowledge of nature which is, in turn, highly influenced. We will attempt to discern the consequences of this implicit philosophy for the analyses of human cognition.

In contrast to the very abstract paradigms which still dominate the foundations of mathematics, physics and biology still constitute themselves respectively around the concepts of *matter* and of *life*, which appear to be very "concrete" although being indefinable (if not negatively) within the internal framework of these disciplines. But they also present the difficulty of essentially having recourse to the requirements of rational coherence, highly mathematized in physics, as well as to the necessities of adequacy, through experimentation, with an independent phenomenality, albeit conceptually constructed.

The specificity of living phenomena will be the object of a relatively new conceptualization, which we will present to the reader by means of a complex play of differentiation and synthesis of physics and biology. Theoretical delimitation is, in our view, a very first step in the construction of scientific knowledge, especially with regard to phenomenalities, so difficultly reducible to one another, which are the living and the inert. Quantum physics provides us with a very important paradigm for understanding this method: along its history, first were constituted the bases of an autonomous theory and objectivity, which were quite different from the well established causal structures of classical (and relativistic) physics and having their own analyses of determination (intrinsic indetermination of the simultaneous measurement of position and momentum; the non-locality and non-separability of certain quantum objects; the absence of trajectories as such within classical space-time ... we will go back to this at length). Then, the problem of unification was brought forth, at least the problem of the construction of conceptual gateways, of relationships, of dualities, and symmetries for objects and theories. Likewise, in our view, the theoretical autonomy of the analyses of living phenomena must precede any attempt, as important as it may be, at unification/correlation to physical phenomenalities.

In this perspective of a highlighting of the theoretical differences as well as the strong correlations, or, of a desirable unity of epistemic nature between physics and biology, it is necessary to articulate and, if possible, to put into correspondence "law of thought" and "law of object", abstract formalization and experience, in the specificity of their different roles within these disciplines. Now, it seems that is one of the first requirements of any natural philosophy to clarify and to interpret, in terms of theory of knowledge, this problematic articulation and, in order to do so, to identify the fundamental principles which ensure its synthesis within these sciences.

Our starting point will be a comparative analysis of the construction of objectivity in mathematics and in physics, two disciplines having been constituted with a strong intertwining with one another at the very origin of modern science. This reciprocal determination goes back at least to Galileo, but in fact, it has its origin within Greek science and extends to the geometrization of physics, which has marked the great turning point of the XXth century, with the movement corresponding to the physicalization of geometry. We will recognize, in this process, Riemannian geometry and relativity, the geometry of dynamical systems and, today, the geometrization of quantum physics.

To this end, our reflection introduces a distinction between "principles of proof" and "principles of construction" which we develop in parallel between the foundations of mathematics and of physics as well as within these disciplines. We will then understand the great theorems of incompleteness of formal systems as a discrepancy (a "gap") between proofs and constructions and we will discuss the so-called incompleteness of quantum physics in order to highlight its analogies and differences with mathematical incompleteness. Our approach lies beyond the debate which has dominated the analyses of the foundations of mathematics for too long, a play between Platonism and formalism, and which has shifted away from any relationship to physics and our life-world. Particularly, we will refer to the cognitive foundations of mathematics, of which the analysis constitutes in our view a primary link to other forms of knowledge, via action within physical time and space and via living phenomena.

The reasons for a centrality of the relationship to space and to time, of the "active access to the world" which accompanies any form of knowledge, will often be emphasized. In this way, the constitution of scientific knowledge will, in our approach, be founded upon these relativizing forms specific to modern physics: the clear enunciation of a reference system and of a measurement, in the broadest sense, is in our view the starting point for any construction of scientific objectivity. Digital modelization, and thereby intelligibility by means of computerization, providing today an essential contribution but not being neutral with regard to this construction, will also be at the center of our analyses.

In this spirit, one of the viewpoints which we will defend here, from the standpoint of physics, consists in seeing in the "geodesic principle" (which concerns "action" within physical or conceptual spaces) one of these fundamental and constitutive principles of knowledge (a "construction principle," in our terms). These principles are not limited to the very strong determination of physical objectivities; they also seem to respond to profound cognitive requirements for any form of knowledge. The way by which the geodesic principle is called to operate within the major theories of physics will lead us to reflect, even more generally, on the roles played by symmetries, symmetry breakings, invariances, and variabilities. In particular, we will closely examine the symmetries proposed and their breaking operated by the computer as a discrete state machine: by these means, in our view, the latter's strength of conceptual framework and of calculus marks the intelligibility of the world, by imposing its own causal structure upon it.

Symmetries, invariances, and their breaking not only manifest through physical phenomenality, but appear to effectively govern its objectivity, at an even deeper level of abstract determination and maybe even of cognitive regulation. It is precisely within this framework that we will examine the relationships that such principles, which contribute to construct and determine scientific knowledge, may bear with their cognitive sources. These relationships constitute modern science in a seemingly contradictory fashion: on the one hand, it could be an issue of mathematical retranscription, adapted to more or less formalized disciplines, of common mental structures of human cognition itself, in such a way that the geodesics, or the symmetries of organisms and of action, could be correlated to their mathematical formulations. On the other hand, it is often an issue of forced highly abstract constructions which are counter-intuitive and which break with spontaneous cognitive representations such as empiricity would tend to develop within the limited framework of daily experience.

We will then attempt to address the relevance of these issues of general invariances for the field of biology and to see the way by which they seem to emerge given the current state of some of biology's sub-disciplines. The situation is actually particularly complex in biology, where the "reduction" to one or another of the current physico-mathematical theories is far from being accomplished (inasmuch as it would be possible or even desirable). In our view, one of the difficulties in doing this lies as much in the specificity of the causal regimes of the physical theories (which, moreover, differ amongst themselves), as in the richness specific to the dynamics of living phenomena. We are thinking here of the intertwining and coupling of levels of organization, of the phenomena of auto-organization and of causal correlation (global and local) which are to be found in biology.

We will thus analyze these aspects as finding themselves in a "boundary" situation for physics. The "physical singularity of living phenomena" refers to the difficulties, which in our view are intrinsic for current physical theories, in grasping these phenomena which enable us to maintain physical states such as those pertaining to life within an extended "criticality" in space and time, in a very specific physical sense, which we will discuss. It is thus physics that enables us to speak of singularities, in a mathematical sense to be clarified (though informally). Yet, the "extended critical states" are not physical as such, or, better, they are not (yet?) addressed by existing theories of the inert. In short, for us, the biological is to be analyzed as a "limit physical situation," an "external limit" with regard to contemporary physical theories. To put it clearly, we have no doubt, or it is our main metaphysical assumption, that we are made just of physical matter, whatever this may mean. However, its peculiar state, the "living state of matter," requires in our view a proper theoretical approach, possibly a *theoretical extension*, in the sense of logic (adding new concepts and principles), of existing theories of the inert. Unification may follow, once a theory of its singularity is clearly established.

This discussion will be conducted all the while explaining the theoretical and conceptual effects of our approaches upon the characterization – or recategorization – of concepts as fundamental as those of *space* and *time* such as they are addressed within the contemporary framework of the natural sciences.

To conclude and to summarize ourselves, in this book we attempt to identify the organizing concepts of some physical and biological phenomena by means of an analysis of the foundations of mathematics and of physics, in the aim of unifying phenomena, of bringing different conceptual universes into dialog. The analysis of the role of "order" and of symmetries in the foundations of mathematics will be linked to the major invariants and principles, among which is the geodesic principle, which govern and confer unity to the various physical theories. Particularly, we will attempt to understand causal structures, a central element of physical intelligibility, in terms of symmetries and their breaking. The importance of the mathematical tool will also be highlighted, enabling us to grasp the differences in the models for physics and biology which are proposed by continuous and discrete mathematics, such as computational simulations.

In the case of biology, being particularly difficult and not as thoroughly examined at a theoretical level, we will propose a "unification by concepts," an attempt which should always precede mathematization. This will constitute an outline for unification also basing itself upon the highlighting of conceptual differences, of complex points of passage, of technical irreducibilities of one field to another. Indeed, a monist point of view such as ours should not make us blind: there is no doubt, in our view, that "physical matter" is unique and that there is nothing else in the universe. Nevertheless, the tools for knowledge which humanity has constructed throughout history in order to make intelligible natural phenomena, are not unified and for good reasons, reasons which are relevant to the very effectiveness of the construction of the scientific objectivity with regard to different phenomenalities. And we cannot claim to unify them by means of a forced methodological and technical monism, according to which such a mathematical method or physical theory, constructed around a very specific phenomenal field, could make us understand everything. Even physics, as a theoretical construction, is far from being unified: quantum mechanics and general relativity are not – yet – unified (the notions of quantum and relativistic field differ). And, as we have mentioned earlier, physicists aim for unification, not reduction, meaning that they aim for a new notion of a field (either of physical objects or of space-time) which will unify them, if necessary by putting each theory into perspective. Such is the case, for instance, with the theory of superstrings which proposes new quantum "objects," or with the radical changes in the very concepts of time and space, such as contemplated by non-commutative geometry. And, since Copernicus and Galileo, Leibniz and Newton – or since always in science – these theories, as for microphysics today, construct their own revolutionary conceptual frameworks, which are in fact counter-intuitive, as well as constructing, in some cases afterwards, their own mathematical tools.

In short, we are monists in what concerns "matter" given that unity can be found, we presume, in this matter which enters into friction with, canalizes and opposes our experiences and theorizing. However, unity is not necessarily to be found within the existing *theories* it may be instead constructed by novel perspectives. The dialog of the disciplines aims to lead us towards new methodological ideas and syntheses: coherence in the intelligibility of the world is a difficult achievement. If it is possible, it is not to be found, we insist, within the claimed unicity and completeness of the theories currently available for addressing one or another of the historically given phenomenal fields. It may be constructed instead by means of conceptual bridges, sometimes of punctual reduction, capable of highlighting points of contact and of friction, by common conceptual foundations. But also by means of theoretical differentiations or dualities and, if possible, of simultaneous changes in perspective within various fields. Conceptualizations which are at the boundary of widespread physical theories are at the center of our attempts in biology.[1]

---

[1] More philosophical reflections as well as technical work, for which this book provides the conceptual frame, may be downloaded from http://www.di.ens.fr/users/longo/

Contents

Preface v

1. Mathematical Concepts and Physical Objects 1
   1.1 On the Foundations of Mathematics. A First Inquiry 7
      1.1.1 Terminological issues? 7
      1.1.2 The genesis of mathematical structures and of their relationships – a few conceptual analogies 10
      1.1.3 Formalization, calculation, meaning, subjectivity 13
      1.1.4 Between cognition and history: Towards new structures of intelligibility 17
   1.2 Mathematical Concepts: A Constructive Approach 19
      1.2.1 Genealogies of concepts 19
      1.2.2 The "transcendent" in physics and in mathematics 23
      1.2.3 Laws, structures, and foundations 31
      1.2.4 Subject and objectivity 37
      1.2.5 From intuitionism to a renewed constructivism 40
   1.3 Regarding Mathematical Concepts and Physical Objects 44
      1.3.1 "Friction" and the determination of physical objects 45
      1.3.2 The absolute and the relative in mathematics and in physics 47
      1.3.3 On the two functions of language within the process of objectification and the construction of mathematical models in physics 48
      1.3.4 From the relativity to reference universes to that of these universes themselves as generators of physical invariances 51

		1.3.5	Physical causality and mathematical symmetry	52
		1.3.6	Towards the "cognitive subject"	55
2.	Incompleteness and Indetermination in Mathematics and Physics			57
	2.1	The Cognitive Foundations of Mathematics: Human Gestures in Proofs and Mathematical Incompleteness of Formalisms		58
		2.1.1	Introduction	58
		2.1.2	Machines, body, and rationality	59
		2.1.3	Ameba, motivity, and signification	61
		2.1.4	The abstract and the symbolic; the rigor	62
		2.1.5	From the Platonist response to action and gesture	65
		2.1.6	Intuition, gestures, and the numeric line	69
		2.1.7	Mathematical incompleteness of formalisms	73
		2.1.8	Iterations and closures on the horizon	75
		2.1.9	Intuition	78
		2.1.10	Body gestures and the "cogito"	82
		2.1.11	Summary and conclusion of part 2.1	83
	2.2	Incompleteness, Uncertainty, and Infinity: Differences and Similarities Between Physics and Mathematics		85
		2.2.1	Completeness/incompleteness in physical theories	85
		2.2.2	Finite/infinite in mathematics and physics	93
3.	Space and Time from Physics to Biology			101
	3.1	An Introduction to the Space and Time of Modern Physics		103
		3.1.1	Taking leave of Laplace	103
		3.1.2	Three types of physical theory: Relativity, quantum physics, and the theory of critical transitions in dynamical systems	105
		3.1.3	Some epistemological remarks	111
	3.2	Towards Biology: Space and Time in the "Field" of Living Systems		113
		3.2.1	The time of life	113
		3.2.2	More on Biological time	115
		3.2.3	Dynamics of the self-constitution of living systems	120
		3.2.4	Morphogenesis	124
		3.2.5	Information and geometric structure	128

		3.3	Spatiotemporal Determination and Biology	132
		3.3.1	Biological aspects	132
		3.3.2	Space: Laws of scaling and of critical behavior. The geometry of biological functions	133
		3.3.3	Three types of time	136
		3.3.4	Epistemological and mathematical aspects	139
		3.3.5	Some philosophy, to conclude	143
4.	Invariances, Symmetries, and Symmetry Breakings			149
	4.1	A Major Structuring Principle of Physics: The Geodesic Principle		149
		4.1.1	The physico-mathematical conceptual frame	151
	4.2	On the Role of Symmetries and of Their Breakings: From Description to Determination		158
		4.2.1	Symmetries, symmetry breaking, and logic	158
		4.2.2	Symmetries, symmetry breaking, and determination of physical reality	161
	4.3	Invariance and Variability in Biology		165
		4.3.1	A few abstract invariances in biology: Homology, analogy, allometry	165
		4.3.2	Comments regarding the relationships between invariances and the conditions of possibility for life	169
	4.4	About the Possible Recategorizations of the Notions of Space and Time under the Current State of the Natural Sciences		175
5.	Causes and Symmetries: The Continuum and the Discrete in Mathematical Modeling			181
	5.1	Causal Structures and Symmetries, in Physics		182
		5.1.1	Symmetries as starting point for intelligibility	186
		5.1.2	Time and causality in physics	187
		5.1.3	Symmetry breaking and fabrics of interaction	190
	5.2	From the Continuum to the Discrete		195
		5.2.1	Computer science and the philosophy of arithmetic	196
		5.2.2	Laplace, digital rounding, and iteration	198
		5.2.3	Iteration and prediction	201
		5.2.4	Rules and the algorithm	203
	5.3	Causalities in Biology		210
		5.3.1	Basic representation	211

		5.3.2	On contingent finality	215
		5.3.3	"Causal" dynamics: Development, maturity, aging, death	216
		5.3.4	Invariants of causal reduction in biology	218
		5.3.5	A few comments and comparisons with physics . .	220
	5.4	Synthesis and Conclusion		220

6. Extended Criticality: The Physical Singularity of Life Phenomena — 225

   6.1 On Singularities and Criticality in Physics 227
      6.1.1 From gas to crystal 227
      6.1.2 From the local to the global 229
      6.1.3 Phase transitions in self-organized criticality and "order for free" 231
   6.2 Life as "Extended Critical Situation" 236
      6.2.1 Extended critical situations: General approaches . 240
      6.2.2 The extended critical situation: A few precisions and complements 242
      6.2.3 More on the relations to autopoiesis 244
      6.2.4 Summary of the characteristics of the extended critical situation 245
   6.3 Integration, Regulation, and Causal Regimes 246
   6.4 Phase Spaces and Their Trajectories 250
   6.5 Another View on Stability and Variability 255
      6.5.1 Biolons as attractors and individual trajectories . 255

7. Randomness and Determination in the Interplay between the Continuum and the Discrete — 259

   7.1 Deterministic Chaos and Mathematical Randomness: The Case of Classical Physics 262
   7.2 The Objectivity of Quantum Randomness 265
      7.2.1 Separability vs non-separability 267
      7.2.2 Possible objections 269
      7.2.3 Final remarks on quantum randomness 273
   7.3 Determination and Continuous Mathematics 274
   7.4 Conclusion: Towards Computability 278

8. Conclusion: Unification and Separation of Theories, or
   the Importance of Negative Results   281
   8.1   Foundational Analysis and Knowledge Construction . . .   281
   8.2   The Importance of Negative Results   285
         8.2.1   Changing frames   289
   8.3   Vitalism and Non-Realism   292
   8.4   End and Opening   297

Bibliography   299

Index   313

Chapter 1

Mathematical Concepts and Physical Objects

Introduction

With this text, we will first of all propose and discuss a distinction, internal to mathematics, between "construction principles" and "proof principles." In short, it will be a question of grasping the difference between the construction of mathematical concepts and structures and the role of proof, more or less formalized. The objective is also to analyze the methods of physics from a similar viewpoint and, from the analogies and differences that we shall bring to attention, to establish a parallel between the foundations of mathematics and the foundations of physics.

When proposing a mathematical structure, for example the integer numbers or the real numbers, the Cartesian space or a Hilbert space, we use a plurality of concepts often stemming from different conceptual experiences: the construction of the integers evokes the generalized successor operation, but at the same time we make sure they are "well-ordered," in space or time. That is, that they form a strictly increasing sequence, with no (backward) descending chains. This apparently obvious property (doesn't it appear so?) yields this well-ordered "line of integer numbers," a non-obvious logical property, yet one we easily "see" within our mental space (can't you see it?). And we construct the rationals, as ratios of integers modulo ratio equivalence, and then the real numbers, as convergent sequences (modulo equiconvergence), for example. The mathematician "sees" this Cantor–Dedekind-styled construction of the continuum, the modern real line and continuum, a remarkable and very difficult mathematical reconstruction of the phenomenal continuum. It is nevertheless not unique: different continua may be more effective for certain applications, albeit that their structures are locally and globally very different, non-isomorphic to this very familiar standard continuum (see Bell, 1998). And this construction is so important that the "objectivity" of real numbers is "all there," it depends solely upon this very construction, based on the well-order of integers, the passage to the quotients (the rationals) and then the audacious limit operation by Cantor (add all limits of converging sequences). One could say as much about the most important set-theoretic constructions, the cumulative hierarchies of sets, the sets constructed from the empty set (a key concept in mathematics) by the iterated exponent operations, and so on. These conceptual constructions therefore obey well-explicated "principles" (of construction, as a matter of fact): add one (the successor), order and take limits in space (thus, iteration, the well-order of the integers, limits of converging sequences).

But how may one grasp the properties of these mathematical structures? How may one "prove them"? The great hypothesis of logicism (Frege) as well as of formalism (Hilbert's program) has been that the logico-formal proof principles could have completely described the properties of the most important mathematical structures. Induction, particularly, as a logical principle (Frege) or as a potentially mechanizable formal rule (Hilbert), should have permitted us to demonstrate all the properties of integers (for Frege, the logic of induction coincided, simply, with the structure of the integers – it should have been "categorical," in modern terms). Now it happens that logico-formal deduction is not even "complete," as we will recall (let's put aside Frege's implicit hypothesis of categoricity); particularly, many of the integers' "concrete" properties elude it. We will evoke the "concrete" results of incompleteness from the last decades: the existence of quite interesting properties, demonstrably realized by the well-ordering of integers, and which formal proof principles are unable to grasp. But that also concerns the fundamental properties of sets, the continuum hypothesis, and of the axiom of choice, for example, demonstrably true within the framework of certain constructions, as shown by Gödel in 1938, or demonstrably false in others (constructed by Cohen in 1964), thus unattainable by the sole means of formal axiomatics and deductions.

To summarize this, the distinction between "construction principles" and "proof principles" shows that theorems of incompleteness prohibit the reduction (theoretical and epistemic) of the former to the latter (or also of semantics – proliferating and generative – to strictly formalizing syntax).

Can we find, this time, and in what concerns the foundations of physics, some relevance to such a distinction? In what would it consist and would it play an epistemologically similar role? Indeed, if the contents and the methods of these two disciplines are eminently different, the fact that mathematics plays a constitutive role for physics should nevertheless allow us to establish some conceptual and epistemological correspondences regarding their respective foundations. This is the question we shall attempt to examine here. To do so, we will try to describe the same level of "construction principles" for mathematics and physics, that of mathematical structures. This level is common to both disciplines, because the mathematical organization of the real world is a constitutive element of all modern physical knowledge (in short, but we will return to this, the constitution of the "physical object" *is* mathematical).

However, the difference becomes very clear at the level of the *proof* principles. The latter are of a logico-formal nature in mathematics, whereas in physics they refer to observation or to experience; shortly, they refer to measurement. This separation is of an epistemic nature and refers, from a historical viewpoint, to the role of logicism (and of formalism) in mathematics and of positivism in physics. We will therefore base ourselves upon the following table:

Disciplines	Mathematics	Physics
1. Construction principles	Mathematical structures and their relationships	
2. Proof principles	Formal/Logical proofs	Experience/observation
Reduction of 1 to 2	Logicism/Formalism	Positivism/empiricism

Let's comment this schema with more detail. The top level corresponds to the construction principles, which have their effectiveness and their translation in the elaboration of mathematical structures as well as in the various relationships they maintain. These structures may be relative to mathematics as such or to the mathematical models which retranscribe, organize, and give rise to physical principles – and by that, partly at least, the phenomena that these principles "legalize" by provoking and often guiding experiments and observation. This community of level between the two disciplines, in what concerns the construction of concepts, does not only come from the constitutive character of mathematics for physics, which we just evoked and which would almost suffice to justify it, but it also allows us to understand the intensity of the theoretical exchanges (and not only the instrumental ones) between these disciplines. Physics certainly obtains elements of generalization, modelization, and generativity from mathematical structures and their relationships, but physics' own developments also suggest and propose to mathematics the construction of novel concepts, of which physics, in some cases, already makes use, without waiting to be rigorously founded. Historical examples abound: be it the case of Leibnizian infinitesimals, which appeared to be so paradoxical at the moment they were introduced – and for a long while after that – and which were never theoretically validated elsewise than by modern non-standard analysis, be it Dirac's "function" which was rigorously dealt with only in the subsequent theory of distributions, be it the case of Feynman's path integrals – which have not yet found a sufficiently general rigorous mathematical treatment, while revealing themselves to be completely operable – or be it the birth of non-commutative geometry inspired by the properties of quantum physics.

The second level, corresponding to that of the proof principles, divides itself into two distinct parts according to whether it concerns mathematics or physics (in that their referents are obviously different). For mathematics, what works as such are the corresponding syntaxes and logico-formal languages which, since Frege, Russell, Hilbert, have been presented as the foundations of mathematics. In fact, the logicism and formalism which have thus developed themselves at the expense of any other approach never stopped to identify the construction principles level with the proof principles level by reducing the first to the second. The incompleteness theorems having shown that this program could not be fulfilled for reasons internal to formalism (they prove that the formal proof principles produce valid but unprovable statements), the paradoxical effect was to completely disjoin one level from the other in the foundations of mathematics, by leading syntax to oppose semantics or, by contrast, by refusing to satisfy oneself with proofs not totally formalized (in the sense of this formalism) as can exist in geometry for example. In fact, it appears, conversely, that, as all of the practice of mathematics demonstrates, it is the coupling and circulation between these two levels that make this articulation between innovative imagination and rigor which characterizes the generativity of mathematics and the stability of its concepts.

Let's now consider physics, where the emergence of invariants (and symmetries) also constitutes a methodological turning point, as well as the constitution of objects and of concepts (see Chapters 4 and 5). But this time, at the level of the proof principles, we no longer find a formal language, but the empiricism of phenomena: experiences, observations, even simulations, validate the theoretical predictions or insights provided by the mathematical models and prove their relevance. As constructed as they may have been by anterior theories and interpretations, it is the physical facts which constitute the referents and the instruments of proof. And there again, a particular philosophical option, related to the stage of development of the discipline and to the requirement of rigor in relation to physical factuality, has played, for the latter, a similar role to that of logicism and especially to that of formalism for mathematics. It consists in the positivism and the radical empiricism which, believing to be able to limit themselves only to "facts," attempted to reduce the level of construction, characterized, namely, by interpretative debates, to that of proof, identified to pure empiricity. The developments of contemporary physics, that of quantum physics particularly, of course, but also that of the theory of dynamical systems, have shown that this position was no longer tenable and that the same paradoxical effect has led, doubtlessly by reaction, to the epistemological disjunction between the levels of conceptual and mathematical construction and of empirical proof (a transposed trace is its opposition between "nominalists" and "realists" in the epistemology of physics). While, there again, all the practice of physicists shows that it is in the coupling and the circulation between these levels that lies the fecundity of the discipline, where empirical practices are rich of theoretical commitments and, conversely, theories are heavily affected by the methods of empirical proofs. And, since for us the analysis of the genesis of concepts is part of foundational analysis, it is this productivity itself that feeds off interactions and which takes root within cognitive processes, which must be analyzed.

It is thus in this sense, summarized by the above schema, despite their very different contents and practices, that the foundations of mathematics and the foundations of physics can be considered as presenting some common structural traits. That is, this distinction between two conceptual instances are qualifiable in both cases as construction principles and as proof principles, and the necessity of their coupling – against their disjunction or conversely, their confusion – is important to also be able to account for the effective practice of researchers in each of these disciplines. Moreover, that they share the same level as for the constitution of mathematical structures characterizing the dynamics of construction principles and feeding off the development of each of them.

If we now briefly address the case of this other discipline of natural sciences which is biology, it appears, in what concerns the structure of its own foundations, to distinguish itself from this schema, though we may consider that it shares with physics the same level of proof principles, that is, the constraint of reference to the empiricity of observation and of experience. However, we are led, at the level of this proof principle, to operate a crucial distinction between what is a matter of *in vivo* (biological as such in that it is integrated and regulated by biological functions), and what is a matter of *in vitro* (and which practically confounds itself with the physico-chemical). But what manifestly changes the most depends, it seems, on two essential factors. On the one hand, the level of what we may call (conceptual) "construction principles" in biology still does not seem well characterized and stabilized (despite models of evolution, autonomy or autopoiesis[1]). On the other hand, it seems that another conceptual level adds itself, one specific to the epistemology of the living, and to which is confronted any reflection in biology and which we may qualify, to use Monod's terminology, as the level of the teleonomic principle. This principle in some way makes the understanding of the living depend not only upon that of its past and current relationships to its relevant environment, but also upon that of the anticipations relative to the future of what this environment will become under the effect of its own activity of living. And this temporality lays itself beside the temporality treated by physical theories. This regulates the physico-chemical action-reaction relation, but, on purely theoretical grounds, we must consider also a biological temporality specific to the organism which manifests itself as the existence and the activity of "biological clocks" which time its functions (see Chapter 3). This conceptual situation then leads us to consider, for biology, the characterization of an extra, specific concept, in interaction with the first two, which we like to call "contingent finality"; meaning by that the regulations induced by the implications of these anticipations, and which themselves open the way to the accounting for "significations" as we will argue below.

As mentioned in the book's introduction, this chapter (and only this chapter) will be based on an explicit distinction of the author's contribution, following the dialog which started this work. The preliminary questions concerning foundational issues in mathematics and physics will then be raised.

---

[1] This is defined as a "process that produces the components that produce the process, ..." typically, in a cell, the metabolic activities are a process of this kind, see Varela (1989), (Bourgine and Stewart, 2004).

1.1 On the Foundations of Mathematics. A First Inquiry
(by Francis Bailly)

1.1.1 Terminological issues?

1.1.1.1 Regarding the term "structure"

There is often confusion, in physics, on the use of terms such as "mathematical structure" and "mathematical formalism": a physicist claims to be "formalizing" or "mathematizing" in a rather polysemic way (which is of course sufficient for his/her everyday work, but not for our foundational analysis of common construction principles). It very well appears that the term "structure" (and its derivatives) in mathematics may be interpreted in two distinct ways, which may be clarified in the light of our previous distinction (proof vs construction principles). The first sense refers to the general usage of the term: a formal mathematical structure characterized by axiomatic determinations and associated rules of deduction, for instance, the formalized structure of the field of real numbers, the structure of numbers defined by the axioms of Peano's arithmetic, the axiomatic structure of transformation groups, etc. The second sense rather refers to a structure as characterized by properties of content more than by formal axiomatic determinations and therefore presents more of a semantic aspect, that is proposed by construction principles (we will detail at length in this book symmetries, ordering principles, etc). As, for example, when we refer to the structure of continuous mathematics or to the connectedness of space. It is in this second sense that Giuseppe Longo and many others seem to employ the term in their criticism of the formalist and set-theoretic approach and it is according the latter framework that we are then led to question ourselves regarding a possible dialectic between the rigidity of an excess of structure and the dispersal entailed by a complete lack of structure.

Using the example of space, will one refer to a sort of space which is completely determined in its topology, its differential properties, in its metric – a space that is very "structured"? Or, conversely, would one refer to a very "unstructured" space, composed of simple sets of points, which enable, in the manner of Cantor, even if that means a loss of all continuity, of all notion of neighborhood, to establish a bijection between the plane and the straight line that is far removed from the first phenomenological intuition of the space in which our body is located and evolves?

We must note, at this stage, that the reference systems used in contemporary physics have recourse to "spaces" presenting somewhat intermediary properties: they do in fact renounce the absoluteness of a very strong if not complete determination (the space of Newtonian and of classical mechanics), but they do not go as far as the complete parcellation as presented by Cantorian sets of independent points. Hence, they conserve an important structure in terms of continuity or connexity all the while losing, via the properties of homogeneity or of isotropy, a number of possible structural determinations. Moreover, invariance by symmetry is sufficiently constrictive or structuring to preclude the definition of an absolute origin, of a privileged direction, or of many other "rigid" properties. Noether's theorem, which we will address in length particularly in Chapters 3 and 5, establishes an essential correlation between these properties of time/space symmetry and the conservation of certain physical magnitudes (energy, kinetic moment, electric charge, ...) that characterize the system, in its profound identity and in its evolution. It appears that it is these properties of symmetry (of invariance) which lead us to characterize the reference space's relevant structure, which is neither too strong nor too weak.

So there would be a sort of theoretical equivalence, in physics, of this type of mathematics where a too strong "structure" would only enable us to construct "isolated" and specific objects, whereas a little bit of structural relaxing would enable us to characterize similarities and to elaborate categories and relations between them.

1.1.1.2 *Concerning the term "foundation"*

The term "foundation" raises similar questions. It appears indeed that the term may be articulated into two quite different concepts.

It could stand for the evocation of an *a priori* origin that is associated with a first intuition, and from which would historically be deployed the theoretical edifice (the phenomenological intuition of continuity or of number, for example) and which would remain as an hermeneutic insistence of the question originally posed. Foundation would then have a genetic status, and it would provide a proof of the hermeneutic relevance of an issue by the ensuing theoretical fecundity. In this sense, the foundation would appear as the basis for any ulterior development or construction. We may ask ourselves if it is not in this epistemological or even genetic sense that we should use the term here.

In contrast, however, and historically, the term "foundation" may also designate an *a posteriori* structure that is quite formal or even completely logicized, and which presents itself as the result of a theoretical evolution and as the very elaborate product of a rational reconstruction which enables us to reinterpret and to restitute all of the concept's anterior work (as in the case of a Hilbertian axiomatic reconstruction, for example). It is partly in these terms that the problematic of foundations was articulated after the crisis at the beginning of the XXth century.

Beyond their contrasting positions regarding the "temporality" of the theoretical/conceptual work (an origin difficult to assign in one case, posteriority never achieved in the other), do these two meanings not underlie different representations of essentiality? As indeterminate as it may still appear although being rich in terms of the ulterior developments it is likely to generate, do we not find different philosophies of knowledge in the very nature of the question which is posed, it being in the first case genetic and in the second case formal? And do we not find differing philosophies of knowledge in the nature of the answer, capable of reinterpreting and of conferring meaning to the effort of which it embodies the outcome and which it summarizes?

Besides these issues, there also arises the question relative to *invariance*: structural invariance and conceptual stability, which Longo makes into one of the most important characteristics of mathematics as a discipline, and which needs to be addressed in length. Now, in the problem at hand, invariance itself seems to take two aspects: on the one hand in the insistence of the original and still relevant question (the question of the continuum, for example, which spurs increasingly profound research), and on the other hand, in the formal structure revealed by research and which, once constructed, presents itself in a quasi atemporal manner (the proof-theoretic invariance related to the validity of Pythagoras' theorem, for example).

Would it not be this which would enable us to explain the double characteristic associated with the concept of foundation but also the double aspect of structure, according to whether it has recourse to a richness founded upon a semantic intuition or to a rigor compelled by formalism?

1.1.2 *The genesis of mathematical structures and of their relationships – a few conceptual analogies*

Still concerning mathematical structures, it is a question here of raising and briefly discussing the issues relating to their genesis and not only to their history. So operating an important distinction between the genesis of mathematical structures themselves and the reconstructed genesis of their relationships, the one and the other appear to stem from different approaches, conceptualizations, and relational processes and to involve distinct cognitive resources.

As for the genesis of mathematical structures, one may distinguish an historical genesis as such which can be retraced and located within a timeline as well as a conceptual genesis of which the temporality is clearly more complex. Naturally, the first is an object of the history of mathematics, in terms of the discoveries and inventions which do not need to be addressed here. The second is of a quite different nature: a given concept or approach having had its time of preponderance is re-proposed by others and then reappears, is developed and is forgotten again until it resurges later (as was the case in physics with the atomic hypothesis, for example, and as is the case with mathematical infinity, which also has had a complex historical specification). Another concept appearing to be autonomous and original, even unique like Euclidean geometry, may finally prove to constitute a particular thematization of a more general trend to which it will be associated from then on and which endows it with a different coloration (this was also often the case in number theory, firstly with the appearance of negative numbers, and then of complex numbers, or was also the case with prime numbers and ideal numbers). The work underlying this evolution is that of the concept, of its delimitation and of its generalization. This work is not linear and unidirectional: it returns to previous definitions and developments, enriches them, modifies them, uses them to generate different ramifications, reunifies them and retranslates them, one into the other. In this way, it internalizes the historical temporality which generated these concepts and makes it into an interpreting and interpretive temporality. As recalled earlier, this is what justifies, with the philosophical approach the author associates with it, the qualification of *formal hermeneutic*, which Salanskis (1991) conferred to it, using as characteristic examples the theories of continua, of infinity, and of space. The approach proposed by Longo, and to which we will return in this book, also consecrates the existence of a hermeneutic dimension to the genesis of mathematical structures, as would suggest the importance he gives to *meaning*, beyond that which is purely syntactic.

The issue of the genesis of *relationships* between structures stems from a quite different problem. This genesis also presents two aspects according to the approach one would favor: the aspect usually characterized as *foundational* and the aspect which we qualify as *relational*.

The foundational aspect corresponds in sum to the formal/set-theoretical approach. It is characterized by the search for the most simple, intuitive or elementary foundations possible, from which mathematics as a whole can be re-elaborated in the manner of an edifice, progressively and deductively, going from the simplest and most elementary to the most complicated or sophisticated. From this point of view, this genesis may be considered as marked by the irreversibility of the process and it is quite naturally associated to a typification of stages. This is supposed – in the first formalist programs – to put into correspondence what we qualify as proof principles on the one hand, and construction principles on the other. Revealing that the former did not coincide with the latter was one of the effects of incompleteness theorems.

In contrast, the relational aspect is more of an intuitional/categorical nature: the structures mutually refer the former to the latter in a network more than they follow one another while overlapping. Interreducibility manifests itself diagrammatically and that which is fundamental lies in the isomorphy of correspondences much more than in a presumed elementarity. There, the mismatch between proof principles and construction principles revealed by the incompleteness theorems, to which Longo will need to return in his response, is no longer really a problem because the issue is no longer to make progression from the foundational coincide with the elaboration of structures. Nor does the recourse to impredicative definitions pose a problem, as we will emphasize, since the network in question is not meant to be conceptually hierarchized in the sense of set theories.

So, just as the foundational genesis – of the set-theoretical type – of the relations between structures evokes the correspondence with a sort of external (logical), one-dimensional and irreversible temporality, relational genesis – of the categorical type – refers to a specific, internal temporality, which is characteristic of the network it contributes to weaving. How could we characterize this characteristic temporality in a way that would not be purely intuitive and which would engage a process of objectification?

Seen from this angle, the genesis of structures and the genesis of relationships between structures, despite their profound differences, appear to form a pair, articulating two distinct temporalities, the one defined somewhat externally and the other derived from an internality (or regulating it). Thus, these temporalities specifically relating to mathematics offer troubling analogies with certain aspects of theoretical biology which itself copes with two types of temporality: the physical temporality of the external relationships of the organism to its environment (which presents all the characteristics of physical time, modulo the solely biological relationships as such between stimulus and response) and the intrinsic temporality of its own iterative rhythms defined not only by dimensional physical magnitudes (seconds, hours, ...) but by pure numbers (number of heart beats over the life of a mammal, number of corresponding breaths, etc). And in the case of the relationships between structures, the parallel may appear to be particularly significant: the irreversible "time" of the foundational derivations associated with set theories echoes the external physical-biological time of the succession of forms of life, in a sort of common logic of *thus, ... hence ...*; while the temporality specific to the categorical relationships of networking rather resonates with the biological time specific to the "biological clocks" (which we will address in Chapter 3) – morphogenesis, genetic activations, physiological functionings – according, this time, to an apparently more restrictive logic in terms of ontological engagement but which is more supple in the opening of possibilities, of the *if ..., then ...* .

Furthermore, it must be noted, curiously (but no doubt fortuitously, given the dynamic and temporalized representations which are often at the origin of the intuitionist and constructivist approaches), the relational construction of the relationships between structures tends to mobilize a semanticity quite akin to the auto-organizational version of the biological theories (Varela, 1989). Indeed, the tolerance relative to impredicativity and self-reference is in tune with the self-organizing (and thus self-referential) approach to the organism (reevaluated relationships of the self and non-self, couplings between life and knowledge, recourse to "looped" recursion, etc). The categorical closure involved with this same constructive approach evokes the organizational closure associated with the "self" paradigm in order to delimitate the identities and qualify the exchanges. Here lies one of the possible sources of the conceptual connections we propose to operate between mathematical foundations and possible theorizations in biology.

In conclusion, if one accepts this analysis, then the term of *construction*, of which the scope proved to be so important in both the epistemology of mathematics and in philosophy (Kant notes that if philosophy proceeds by means of concepts, mathematics proceeds on its part by means of the construction of concepts), is both destabilized and enriched. In particular, it is brought to bear two distinct meanings which do not mutually reduce themselves to one another.

On the one hand, indeed, we would have a construction that is irreversible in a way (at least *a posteriori*) which leads from infrastructures to superstructures: the construction of ordinals from the empty set, for example, or of rational and real numbers from natural numbers, etc. as presented by the formalist/set-theoretical approach. And on the other hand, we would have a construction that is much closer to that which extends to characterize intuitionism and constructivism as such, which principally concerns the relationships between mathematical structures as presented by category theory: not only from structural genesis to structures but also restitution of the effective processes of constitution. Integer numbers, we will claim, are constructed and grounded on the manifold experiences of ordering and sequencing, both in space and time: the intuition of the discrete sequence of a time moment, for example, is at the core of the intuitionistic foundation of mathematics. Yet, the constituted invariant, the concept, we will argue, requires many active experiences to attain the objective status of maximally stable intersubjective knowledge.

The role of time in the construction of knowledge will lead us to raise the issue of the resulting concepts of temporality, by attempting to put them into relationship with the concepts of temporality that can be found in physics, but also in other disciplinary fields, namely in biology. More generally, these considerations will lead to questions relative to putting into perspective the term of "conceptual construction" and to the delimitations of the meanings the latter may bear in distinct scientific situations.

1.1.3 *Formalization, calculation, meaning, subjectivity*

1.1.3.1 *About "formalization"*

The first question still seems to concern terminological aspects, but its clarification may have a greater epistemological dimension. In many disciplines, notably in physics, the term "formalization" (and its derivatives) is virtually equivalent to that of "mathematization" (or, more restrictively, of "modelization"). This is visibly not the case in mathematics and logic, where this term takes a meaning which is much stronger and much more defined. Indeed, in the tradition of prevailing foundational programs, this term is used in a strictly formalist sense which is, moreover, resolutely finitary. It is not surprising that the term "formalizable" is almost synonymous with "mechanizable" or "algorithmizable." How then may we qualify other "formalizations" which, all the while remaining within the framework of logic, do not really conform to these very constraining norms, while nevertheless presenting the same rigor in terms of reasoning and proof? If, as observed by Longo (2002) the notion of proof in mathematics, as opposed to Hilbertian certitude, is not necessarily decidable (a consequence of Gödelian incompleteness, to which we will return), then, what will explicitly be, if this is possible to explain, the objective criteria (or at least those which are shared consensually, intersubjectively) that will enable the validation of a deduction?

Moreover, in a somewhat similar order of ideas in what distinguishes the formal from the calculable, how is the situation, regarding the issue of infinity, which appears in non-standard analysis where it is possible to have formally finite sets (in that they are not equipotent to any of their own parts) that are, however, calculably infinite, in that they comprise, for instance, infinitely large integers? Would it be an abusive use of the concept of "formal" (or of "calculable")? Is there an abusive mixture between distinct logical types, or is it of no consequence whatsoever? Would it suffice to redefine the terms? Besides, concerning the reference to (and use of) "actual infinity" in mathematics, questions arise regarding Longo's precise stance, a position which we would like to develop in this text. His intuitionist and constructivist references seem to lead to a necessity to eliminate this concept, whereas he appears to validate it in its existence and its usage within mathematical structures as in proofs.

1.1.3.2 *Regarding the status of "calculation"*

In what concerns the issue of "calculation", let's note that in physics, each calculation step associated with the underlying mathematical model does not necessarily have an "external" correlate, in physical objectivity (moreover, it is possible to go from the premises to the same results by means of very different calculations). Conversely, it would appear that from a mathematically specific point of view, each stage of a calculation must have an "external" correlate (external to the calculation as such), that is the rules of logic and reasoning which authorize it.

Similarly, mathematical "inputs" (axiomatic, for example) and the corresponding "outputs" (theorems) seem relatively arbitrary (they only satisfy the implication of the *"if ... then ... "*), whereas physical "inputs" (the principles), like the outputs (observational or experimental predictions) are narrowly constrained by the physical objectivity and phenomenality which constrain the mathematical model.

Hence the following questions: how and where does the construction of meaning in mathematics occur, as we present it with regards to calculation? What about the "significations" associated with the very rules which regulate it?

1.1.3.3 *On the independence of certain results and the role of significations*

Apart from many other reasons, it is necessary to relate here (see also Chapter 2) results of incompleteness and undecidability in order to produce a critique of the formalist program and of the approach it induces with regard to structure (strictly of a syntactic nature) and to the absence of signification of proof. Likewise for the independence of the Continuum Hypothesis (the impossibility of constructing cardinalities between the countable infinity and the continuum of the real numbers), to which Longo confers somewhat the same critical role, but rather in relationship to formal set theory. However, one can only observe that the history of geometry is marked by similar problems, for instance, with the fact that after centuries of unfruitful research of the independence of Euclid's fifth axiom of which the negation opens the way to non-Euclidean geometries. Hence the question: would we make the same sort of critique regarding the geometric axiomatic, or would we consider that it is in fact a different approach inasmuch as it directly involves "significations" (and in what way)?

But we do know, on the other hand, that such significations mainly based on perception or on language habits may be misleading (illusions, language effects, ...). So how may we preserve what we gain from taking distances from a sort of spontaneous semantic, all the while conserving the dimensions of "meaning"? Would we not be brought to relativize this "meaning" itself, to make it a tributary of the genesis of mathematical structures and proof, to historicize it, in the way that mathematical (or physical) intuition historicizes itself in close relationship with the conceptual evolution of which the forms, content, and avenues change in function of the knowledge cumulated?

1.1.3.4 *On the relationships between rationality, affect, and objectivity*

If all do agree to acknowledge the importance and operativity of affects and emotions for scientific creativity and imaginativeness, is there not, however, a double stake of separation between the two, at the two opposite poles of the scientific approach? At the origin of this approach, including from an historical standpoint, the separation would be between the rational approach and magic (or myth), in the view of achieving objectivity and rationality. Following the process of creation (or of comprehension, that is, in a certain sense, of re-creation), the separation would be between the singular specificity of the subject involved by emotions and the constraint of the communication necessary to the establishment of an intersubjectivity, which would enable the construction of objectivity.

The excess of formalism, from this standpoint, would have been to confuse the condition which enables this latter distinction with the elimination of the significations themselves by seeking to reduce the objective construction to a play of "pure" syntax. In this regard, the formalist approach is not exempt of an interesting historical paradox in that it was initially conceived in relationship to an indubitable ethical dimension, very rich in significations. Leibniz, in his search of a universal characteristic explicitly had as one of his objectives the elimination of violence from human inter-relationships, the *calculemus* being meant to replace power struggles, by the very fact that subjective interests would be eliminated. And this concern – if not its effective actualization in the form of logicism – should be authenticated. In this sense, the only alternative would indeed appear to reside in the construction and cognitive determination of these *cognitive invariants* such as evoked by Longo.

But do we currently have the capacity to identify the invariants (of a cognitive nature and no longer only disciplinary) which would enable for their part to ensure this construction of objectivity in a way that would preserve these significations (or at least, in a way which would preserve its generativity) without, however, including singular idiosyncrasies? The effective practice of many disciplines – including mathematics where the referent is purely conceptual – does seem to indicate that this should be possible. Yet, a more specific and general statement of the ins and outs of these cognitive invariants remains much more problematic.

1.1.4 *Between cognition and history: Towards new structures of intelligibility*

The attempt to analyze processes of construction of knowledge and of constitution of scientific objects, which we will discuss here, does not exactly fall under the framework of the cognitive sciences, nor *a fortiori* within those of the neurosciences. It is rather to be considered as an approach which we could call "complementary," all the while referring often to human cognition. We will do this by analyzing the foundations of mathematics and the steps governing the construction of scientific objectivity and knowledge in their contemporary dynamism inherited from a history clarifying its sources.

First, the development of mathematics, then of physics, and of biology today, presents characteristics, which are sometimes similar, but often different, and which illustrate our cognitive capacities and their operativity in fields which stand among the most advanced; thus, the analysis of these developments provides an area of study and reflection which seems particularly adapted to our analyses and to the conceptual frameworks which we propose.

The history of sciences such as those being conducted today demonstrate quite well, during the stages of their deployment, the extent to which, by means of various evocative metaphors as well as by thorough inquiries, these disciplines have served as a source of inspiration and as examples for the investigation of nature and of the modes of functioning of human cognition. Not that everything of the human would resemble or conform to this: far from it, actually, because there is for instance a lack of the other dimensions of a specifically ethical or artistic nature (although a certain form of estheticism of theories will be present and have its effect on the scientist), but the scientific domain presents itself as the locus of a major advance in terms of human rationality and we believe that this is called to be reflected in most human activities within society.

So this was for the global perspective. Regarding the more specific contents, if we want to be both rigorous and operative in our approach, we cannot make as if they escaped from their constitutive history or to the interpretive controversies which they have caused and continue to cause. We may recall foundational crises in mathematics and formalist drifts, foundational crises in physics and the emergence of very novel and counter-intuitive theories that are subject to conflicts of interpretation, the eruption of biology as a new and expanding field, which is, however, threatened by the reductionism of the "all genetic" of which the success are nevertheless obvious, etc. All these aspects will therefore be the more present in our discussions inasmuch as they enable us to better understand the stages to which these disciplines have developed and that their trace remains within the very content of the theories. As there is no *plausible epistemology* without a quasi-technical mastery and quasi-internal analysis of the scientific contents, we believe that there is no plausible approach to the conditions of development and the properties of operativity of human cognition without a thorough investigation of the fields where it has been deployed in a privileged fashion.

1.1.4.1 *Memory and forgetfulness in mathematics*

An example of this paradigmatic play between human cognition and history is the reference we will make to "memory." We will return to this in Chapter 2 and address the issue of mathematical temporality via that of memory (of concepts, of the methods of its elaboration, its transformations, etc). For now, let's emphasize that we are proposing here a venue aiming to identify the characteristics of this specific temporality which is related to the construction of mathematical objectivity and which manifests in the genesis of its structures. What this approach suggests, with regard to significations, appears to be quite interesting and convincing. We will also address, but much more briefly, the issue of forgetfulness in the constitution of conceptual and mathematical invariants, namely by invoking the relationships between consciousness and unconsciousness. It is, however, in regard to this aspect of forgetfulness (and of a certain form of atemporality which appears to be related to it) that an apparently important question may be posed. Now, as constituents of the invariance and conceptual stability which characterize the mathematics often highlighted by Longo, memory and forgetfulness are to be understood as cognitive phenomena, specific to the human individual all the way from his animality to the communicating community that characterizes his humanity, but they must also and therefore be understood as historical phenomena. Forgetfulness of "that which is not important" (relative to a point of view, an objective, an intention) and selective memory of "that which is relevant" contributes in constituting the objectivity and very object of knowledge itself, as an invariant of manifold active experiences.

The culminant point of forgetfulness and of atemporality, in what concerns the foundations of mathematics, seems to have been reached with the formalist/logicist program and the elimination of the meanings which it explicitly advocated. This approach did obliterate memory, but intended on the other hand to offer the advantage of complete communication and of an alleged universality (independently of distinct cultures, of particular traditions, of subjective singularities and thus, in its spirit, independently of sterile confrontations as well as of conceptual ambiguities). Without returning to previous questions, how may a revelation of these cognitive and historical invariants that are still theoretically problematic, but apparently empirically proven, enable us to articulate the plays and interactions of this necessary constitutive memory and of this forgetfulness which is also constructive? How would it enable us to articulate this temporality specific to mathematical genesis (or even to scientific concepts) and this coefficient of atemporality enabling both intercultural validation and accumulation?

1.2 Mathematical Concepts: A Constructive Approach
(by Giuseppe Longo)[2]

1.2.1 *Genealogies of concepts*

Let's more closely tackle now the idea of a parallel between the constitution of mathematical concepts and of physical objects. We will only be able to respond partially to this inquiry and shall rather reflect upon the meaning of the relativizing constructions specific to mathematics and to physics, within an explicative and foundational framework inspired by the arisen questions. We already hinted at the identification of mathematical and physical "construction principles." But our project is wider, because it is a question of grounding the two "constitutive histories" within our worldly living being, to grasp this biological and historical "cognitive subject," which we share and which guarantees us the objectivity of our forms of knowledge. It is not a question of unifying by force the epistemologies of differing disciplines, but to make them "exchange between themselves," to reveal the reciprocal dependencies, the several common roots. The analysis we propose here will thus base itself upon the following principles:

- The problem of the foundations of mathematics is (also) an epistemological problem.

---

[2] This section by Longo, jointly with the Introduction above, appeared in **Rediscovering Phenomenology** (L. Boi, P. Kerszberg, F. Patras ed.), Kluwer, 2005.

- Any epistemology (of mathematics) must refer to a conceptual genesis, as a "process of construction of knowledge."
- The epistemology of mathematics is an integral part of the epistemology of the sciences (the exact sciences, at least).
- A constitutive element of our scientific knowledge is the relationship, established in the different sciences, to space and to time.

In short, a sensible epistemology of mathematics must try to explicate a "philosophy of nature," a term which is dear to the great minds of the XIXth century. As it is, mathematics is one of the pillars of our forms of knowledge, it helps to constitute the objects and the objectivity as such of knowledge (exact knowledge), because it is the locus where "thought stabilizes itself"; by this device, its foundation "blends" itself to other forms knowledge and to their foundations. Moreover, the conceptual stability of mathematics, its relative simplicity (it can be profound all the while basing itself upon stable and elementary, sometimes quite simple, principles) can provide the connection which we are looking for with the elementary cognitive processes, those which reflect some of the world's regularities in our active presence within that same world, as living beings (and living in intersubjectivity and in history). For these same reasons, the theories of knowledge, from Plato to Descartes, to Kant, Husserl or Wittgenstein, have all addressed the question of the foundations of mathematics, this "purified knowledge," both mysterious and simple, where notions of "truth" and of "proof" (reasoning) are posed with extreme clarity. The problem of the cognitive foundations of mathematics must therefore be analyzed as an essential component of the analysis of human cognition. Within that framework, we will attempt to analyze in what sense "foundations" and "genesis" (cognitive and historic) are strictly related. The very notion of "cognitive foundations" explicitly juxtaposes foundations and genesis.

In this study, the notions of time and space, which we use, do not refer to "natural entities," but rather to the play between sensible experience and conceptual frameworks which allow the natural sciences to manifest themselves. That was in fact the inquiry of the great geometers (Riemann, Helmholtz, Poincaré, Enriques, Weyl, ...) who tried to pose the problem of the foundations of mathematics within the framework of a philosophy of nature. But the analysis, which came to dominate afterwards, stemmed from a very clear division between logical (or formal) foundations and epistemological problems, particularly those presented under the form of this relationship to time and space which ground mathematics in this world.

Frege explicitly denounces the "delirious" situation in which the problem of space finds itself, because of the emergence of non-Euclidean geometries (Frege, 1884), and proposes a "royal way out," by laying the bases of a new discipline, mathematical logic. Mathematics itself is the development of "absolute laws of thought," logical rules outside of this world and independent of any cognitive subject. For that, Frege introduces a very clear distinction between "foundations" and "genesis," he breaks any epistemological ambition, all the while attacking "psychologism" (as of Herbart/Riemann) and "empiricism" (as of John Stuart Mill). The former try to understand which "hypotheses" (which "*a priori*") allow us to make physical space (and time) intelligible to the knowing subject, while the latter relates mathematics to a theory, alas too naive, of perception. Faced with all these first attempts at a "cognitive analysis" of mathematics, Frege proposes a philosophy centered upon a very inflexible dogma, the logicist dogma, according to which mathematics has no psychologico-historical or empirical genesis. It is, according to him, a constituted knowledge, concepts without conceptors. This philosophy, this dogma, is at the origin of the fundamental split, which will accompany all of the XXth century, between foundational analysis and epistemological problems, between mathematics and this very world it organizes and makes intelligible.[3]

Moreover, for Frege, geometry itself, as given by numerical ratios (Frege, 1884), bases itself on arithmetics; and the latter is but the expression of logical laws, because the concept of number is a logical concept and induction, a key rule of arithmetics, is a logical rule. Finally, the continuum, this difficult stake of phenomenal time and space, is also very well mathematized, in Cantor-Dedekind style, from arithmetics.

So there are the problems of time and space and of their mathematization, neglected to the benefit of their indirect foundation, via arithmetic, upon logic; pure concepts, with no relationship whatsoever to sensible experience nor to physical construction. Conversely, this relationship was at the center of the inquiry of the inventors of non-Euclidean geometries: Gauss, Lobatchevsky or Riemann did not play the logical negation of Euclid's fifth axiom and of its formal developments, but they proposed a "new physics,"

---

[3]For us, however, the "almighty dogma of the severance of principle between epistemological elucidation and historical explicitation as well as psychological explicitation within the sciences of the mind, of the rift between epistemological origin and genetic origin; this dogma, inasmuch as we do not inadmissibly limit, as it is often the case, the concepts of "history", of "historical explicitation" and of "genesis", this dogma is turned heads over heels" (Husserl, 1933: p.201).

a different organization of the world (see Lobachevskij, 1856; Riemann, 1854). It also happens that the numerical relationships may possibly found Euclidean geometry, but surely not other geometries, because Euclidean geometry is the only one which preserves these relationships (it is the only one whose group of transformations – of automorphisms – which defines it, contains the homotheties[4]).

Now it is doubtless that mathematics has a logical as well as a formal foundation (a distinction will need to be made here), but it is in fact a "three-dimensional" construction. It constitute itself within the interactions of the logical and totally essential "if ... then" (first dimension), of perfectly formal, even mechanic calculus (second dimension), but also in a third conceptual dimension, these constructions of (and in) time and space, which mingle it, even more so than the two others, with the different forms of knowledge. And the epistemological problem then poses itself as an analysis of the constitution of the invariants of language and of proof, these invariants which we call "logic" and "formal systems," as well as the invariants of time, and of space, upon which we construct our geometries, these "human constructs ... in our spaces of humanity" as Husserl says in the "Origin of Geometry" (see below). The problem is thus posed from the analysis of this very peculiar form of knowledge which is mathematics, from its cognitive roots, be they pre-human, to its communicable display, with its thousands of mediating levels.

Axiomatic conventions and logico-formal proof are actually but the ultimate results of a constitution of meaning, common notations of concepts rooted in "our living practices," to put it as Wittgenstein would do, in our "acts of experience" (Weyl): logico-formal analysis is a necessary accompaniment to this latter part of the epistemological process, the analysis of proof, of certain proofs, but it is insufficient (it is essentially "incomplete," some theorems tell us). The foundational analyses of mathematics must thus be extended from the study of deduction and of axiomatics to that

---

[4]Hilbert, as a great mathematician, will manage quite well otherwise. Thanks to the Beltrami-Klein interpretation of non-Euclidean geometries within Euclidean geometry, he will give a correct immersion (interpretation) of his axioms for geometry within arithmetic via analysis (Hilbert, 1899). But, for the latter, he will not look for a "logical meaning," unlike Frege. Indeed, once geometry (ies) is (are) interpreted within arithmetic, a finitary proof of its (thus, their) coherence (of non-contradiction) would suffice for its foundational analysis, entirely and exclusively centered around its problem of coherence; its a pity it doesn't work, because arithmetic does not have, itself, any arithmetic (finitary) proof of coherence. To the contrary, we manage with an infinite piling of infinities, or by proof founded upon geometrical judgments, see Chapter 2.

of the constitution of concepts and of structures; but this is impossible without a parallel analysis of the constitution of the physical object and of perception.

1.2.2 The "transcendent" in physics and in mathematics

There is no doubt that there exists a reality beyond ourselves, which enters into "friction" with our actions upon it and which, moreover, "canalizes" them. Husserl uses a word from the idealist tradition to designate this reality: he considers the notion of transcendence. In a very common interpretation of this word, and quite independently from Husserl, the following deduction is usually made, first in physics, then in mathematics: the "properties" of the world (physical, numerical, mathematical, ...) are transcendent and, moreover, are not all known. They are therefore "already there," they pre-exist. The objects of the world around us have well-established properties that are quite stable and invariant in relation to our senses: I look at this pencil, I touch it, even its odour confirms its "objectivity," independently of the specific sense I use to explore it ..., it is thus already there, it pre-exists my explorations, with all its properties. In a completely analogous manner, the properties of numbers, of mathematical structures do not depend on notation (for numbers: decimal, binary ...) nor on other details of representation, of the mathematician exploring them ... therefore they pre-exist.

Now it is the word "property" – in physics, in mathematics – that must first be agreed upon: a property is "talked about," it is first of all an expression in these languages through which we try to speak of the world, to organize it and to give it meaning, a meaning shared with others. But the world canalizes our efforts to obtain knowledge and displays some resistance (causes friction) to our propositions to organize it. "Properties," as we render them through intersubjectivity by words, are not in themselves isomorphic to absolute facts that are "already there," possibly well established or that would manifest themselves under well established forms of linguistic structures; by our active gaze, in our exchange with others, we propose a structure with hints of a reality which is there, as unorganized frictional matter. Thus, through language, pictures, gesture, we unify certain phenomena, we draw contours upon a phenomenal veil, which is an interface between the world and us. The transcendent is constituted, it is the result of a constitutive activity, of a process which precedes the individual or that the individual performs mostly with others. This process is best synthesized as the result of a *transcendental* (and not transcendent) activity, and such is the lesson we draw from Husserl.

It is no coincidence if the many examples of objects proposed by ontologizing philosophies, in mathematics, in physics, refer to medium size manufactured objects, all the while attempting to escape the problem of cognitive relativism. These thinkers of ontology, of essences, rarely refer to the "objects" of quantum physics, for example, in order to propose an ontology that is much more difficult to take on, of the electron, of the photon ... But even these medium size manufactured objects, of an apparently such simple ontology, if it is true that they are really there, are just as much constituted as the concept they are associated with. The pencil is constructed, in history, at the same time as the concept of the pencil. Both are related to drawing, to writing, as human activities. They are pre-existent, the object and the concept, for the individual subject, they are not so for humanity, in its history. There was no pencil, nor table, nor a pot such as the one laid on Kurt Gödel's table, before the beginning of our human acting and thinking. On the other hand, there was surely already a physical "reality" (for Galileo, less so for Tales), but its organization and its interpretation as photon, electron ... robust, stable, in fact mathematical, was not yet there, nor was it's organiation into pots, pencils, and tables before the blossoming of our humanity. And this approach, we think, does not face the dangers of relativism, because the objectivity of the constructed, of the concept, of the object, lies in the constitutive process, which is itself objective.

Cassirer, quoted by Parrini in a work whose goal is to overcome the fracture between absolutism and relativism, partially addresses this theme (Parrini 1995, p.118): "if we determine the object not as absolute substance beyond all knowledge, but as object which takes form within the progression of knowledge itself," then, "this object, from the viewpoint of the psychological individual, can be said to be transcendent," despite that "from the viewpoint of logic and of its supreme principles," it must "be considered as immanent." Ideality, the concept as "conceived," "a cut-out" ("decoupage") performed upon the world in order to give it contours, to structure it, will thus detach itself from subjective representation, despite that it may have its origins within the community of subjects, in what they share: similar bodies and brains from the start, in the same world, and all that which they build in common, in their common history. It is thus not a question of writing a history of individuals, but of tracing back the origin of an idea; not historicizing relativism, but a reference to history as an explication of our "being together in the world," the locus of the active constitution of all our forms of knowledge.

In the case of the objects of physics, of microphysics in particular, this activity of the construction of objects by "conceptual carving" is rather clear: electrons, muons, fermions, quantum fields ... are not already there. They are concepts that are proposed in order to unify, to organize, to understand the signals the world sends us. These signals are not arbitrary and they are also the result of an active exploration. In order to obtain them it was necessary to develop rather complex measurement instruments, which are themselves the result of a theory. All the instruments for physical measurement, and more so those of microphysics, are constructed after an enormous theoretical commitment: I want to measure this but not that, by using these materials but not other ones, I "look" here and not there. The "facts" which result from this, as Goodman would say, are thus "small-scale theories" themselves.

Let's consider for example the wave–particle duality in quantum physics. The photon, the electron, present themselves as "waves" or "particles" depending upon the experimental context: specific instruments are put into place, in fact the experiment is prepared from the viewpoint of a certain theory .... The object that will result from this will depend as much upon the theoretico-experimental framework as it will upon friction – "the canalization of thought" that nature imposes upon and through these tools. A certain viewpoint will show us the particle, another will show us the wave. More precisely, we will obtain macroscopical properties on a screen, on a detecting device, and by a process just as important, we will interpret them as symptoms of the "existence" of a particle or of a wave. There is no duality as such for the physical object, but a context of reconstruction of the world where we are as present as the object under observation.

Properties, then, are the "explicated" result of an organizing of clues, of a group of facts, which are themselves "little theories." But reality is there, doubtlessly, because it canalizes our efforts to obtain knowledge in non-arbitrary directions, it causes friction, by opposing itself to our theoretical propositions, great and small, these "properties" spoken of in our languages. The transcendency of these properties, as if they were already constituted, as "ontologies," is a "*flatus vocis*" to which we contrapose the constitutive process of the transcendental which is at the center of Husserl's philosophy. It is our task, when referring to different forms of scientific knowledge, to enrich and to specify this so very fuzzy word, the notion of "property" for the physical world, as well as that of mathematical property.

1.2.2.1 Transcendence vs transcendental constitution: Gödel vs Husserl

So let's move on to mathematics. In this discussion we refer to one of the most interesting among thinkers having an "ontologizing" tendency (and one of the greatest mathematicians of the XXth century), K. Gödel. Actually, Gödel also proposes a strict parallel between physical objects and mathematical concepts, although from a perspective different from ours (the similarity of "ontologies" or of "independent existence"): "It seems to me that the assumption of [mathematical] objects is quite as legitimate as the assumption of physical bodies and there is quite as much reason to believe in their existence" (Gödel, 1944) ... "the properties of these concepts are something quite as objective and independent of our choice as physical properties of matter ... since we can create [them] as little as the constituent properties of matter" (Gödel, 1947). So, physical bodies, and constituent properties of matter, as well as mathematical concepts are all preconstituted entities, possibly the ultimate building blocks, independent of or transcending the cognitive subject (not "created"). Again, even the word property, as referring to outside objective states of affairs, is used in a naive, ordinary way, even for constituent elements, it seems, whose analysis belongs to the entangled constructions of microphysics, where the constitutive polarity "subject/object" is at the core of the modern perspective in quantum mechanics (indeed, since the 1930s).

In his masterpiece about the foundations of mathematics, "The Origin of Geometry," Husserl frequently emphasizes the role of the transcendental constitution of mathematical objects. The epistemological problem they pose is, for him, a "problem of genesis," a "historical problem" (see the footnote above). Geometry, as an attempt (and mankind makes many) to make space intelligible is the result of an activity by "our communicating community"; it is "the constituted," the result of a non-arbitrary process, which grounds our constitutive hypotheses within certain regularities of the world, regularities, "donations" which impose themselves upon us; these regularities are themselves "already there" (the connectivity of space, isotropy, symmetries – inspiring ourselves by Riemann and Weyl). But it is us who choose to see them.

I have a Jovian friend who has five legs, three and a half eyes and no, absolutely no, symmetry to his body. He sees not or does not give any importance to the symmetries of light reflected by a surface, or to crystals, for example, these symmetries which are before our eyes, before his eyes;

and his mathematical structures are not imbued with symmetries like ours (from Greek geometry of the dualities and adjunctions so well described in the Theory of Categories). They are rather constructed around "zurabs," an essential regularity from his perspective, but which we do not see or which we neglect. It goes likewise for colors; he sees a bandwidth beyond violet, where one can find, as a matter of fact, splendid colors. He, therefore, cannot appreciate this marvelous human construction, rich in history, that we call "painting": Titian's colors are invisible for him. Just like we do not see his masterpieces, of such beautiful ultraviolet colors.

The two constructions are not arbitrary, light's waves (or the reality we categorize as such) "are there," just as are the symmetries of crystals or of light bounces, but our active presence interacts with these elements of reality in order to choose, emphasize, correlate some of them, but not others, to gives names, not arbitrary names because they are rich in history and in meaning, to certain color bandwidths and not to others. Moreover, our action interpolates the missing elements, proposes links by analogies derived from other experiences; it integrates a variety of acts of experience in order to create a new structure, an inexisting network between "the things" of the world. To figure out, among the regularities of the world and among the foundational acts of any form of knowledge, which ones are at the origin of mathematics, is one of the tasks of the analysis of the cognitive foundations of mathematics. Husserlian phenomenal analysis may be one tool, if we do not limit ourselves to a fuzzy notion of "transcendence," but if we recover the richness of "transcendental constitution," as we did. Unfortunately, most anti-formalist mathematicians, and even the greatest of mathematical logic, such as Frege and Gödel, insist upon the "transcendent" ("the properties and the objects of mathematics pre-exist, just as do the properties and the objects of physics"). In fact, Gödel, while knowing Husserl, does not refer to the "genesis," to the "history" (in the sense of Husserl (1933) of this constitution at the center of our conceptual constructions.[5] Gödel thus

---

[5] See the discussions reproduced in Wang (1987). Follesdal (1999) makes the generous effort of reading some of Husserl in Gödel. However, for Gödel, the existence of mathematical objects is as external to us as that of physical objects, in that both types pre-exist: "they are independent of our definitions and our constructions"; the intuition of mathematical objects (sets, actually) is a form of "physical perception" (Gödel, 1944, supplement in 1964), in the most naive sense of the term "perception," a sensorial "input" that reaches us as -is. We will hint to the profoundness of Poincaré's sketch of a theory of perception, for example, where we find a true attempt at epistemology in mathematics, rooted in a "philosophy of nature." In the many papers that had a major influence on the contemporary philosophy of mathematics, there is nothing but transcendence in Gödel, even in the quotations chosen by Follesdal, without all the remainder of the phenome-

remains, in mathematics and in physics, at a stage of a realism, which neither specifies the notion of property nor that of object: it has only the objects and the properties derived "from sensations," properties of a physics of "medium size objects" (this table, a pencil ...), a physics which no longer exists, decades after work and debate in relativity, in the physics of critical systems and in quantum physics. The failure of this "realist" epistemology of mathematics is parallel to the absence of an epistemology of physics.

It should be clear though that we have been mainly discussing of Gödel's "realist" position, not only as a tribute to the mathematician (of whom the work on types, in 1958, as well as that on recursion and incompleteness, in 1931, made its mark on XXth century mathematical logic, as well as on the work of this author), but also because his philosophy is by far the most profound among philosophies of mathematical "realism/Platonism." Alain Badiou (Badiou, 1990) emphasizes the richness of this Platonism, alone, in mathematics, resembling that of Plato: *thought envelops* the object, while the idea *is* "already there," but as the name of that which is thought and which would remain unthinkable if not activated within thought. Moreover, for Gödel, as we are reminded, "the objective existence of the objects of mathematical intuition ... is an exact replica of the question of the objective existence of the outside world" (Gödel, 1947). This approach, all the while bringing the question of a mathematical ontology closer to that of an ontology of physics, is far more promising than the realism common in mathematics, a funny mix of vulgar empiricism and of idealism, with the worst shortcomings of each of these two philosophies. However, the difference, relative to the approach sketched here, is given by the understanding of the object as constituted; it is not the existence of physical objects or of mathematical concepts that is at stake, but their constitution, as *their objectivity is entirely in their constitutive path*. It is thus necessary to take Gödel's philosophy, for what it puts into mathematical and physical relation, and to turn it head over heels, to bring it back to earth: one must not start "from above," from objects, as being already constituted (existing), but from the constitutive process of these objects and

---

nal analysis characterizing Husserl; transcendence without transcendental constitution, as a constitutive process of knowledge, without this "Ego" that is co-constituted with the world, which is at the center of Husserlian philosophy, particularly of its maturity. Gödel's late and unpublished writings seem to broaden these views and account for an approach carrying more attention to constitutive paths; yet, a philosophy matters also for the role it has had in history.

concepts. This requires a non-naive analysis of the object and of physical objectivity, as well as a non-passive theory of perception.

1.2.2.2 *Conceptual constructions: history vs games*

To summarize, the objects of mathematics are "outside of ourselves" (transcendent) only as much as they belong to a constituted, which precedes our subject: they are a co-constituted, at the same time as the very intelligibility of the world, by our "living and communicating community." They are not arbitrary because they are rooted in the regularities of reality, to which are confronted our living beings in the world. They are (relative) invariants, first, of time and space, that we then develop by constructing a whole universe derived from conceptual structures, with the most stable tools of our understanding, these invariants of language and of intersubjectivity that we call "logic" and "formalisms": these as well are *the result of a praxis*, the practice of human reasoning, beginning with the Greek agora, in human interaction. In this sense of a previous phylogenetic and historic constitution of their construction principles, and not any another, the objects of mathematics may have properties of which "we do not know," as not yet engendered properties within a more or less precisely given conceptual universe. Take the integers, for example. Once presented, by 0 and the successor operation, as the mental construct of an infinite sequence, discrete and well-ordered (you can picture it, aligned from left to right in a mental space, right?), we can surely give ourselves a language (that of Peano-Dedekind, for example) and enounce an infinity of properties for the elements of this sequence which "we do not know." We will then need to exercise some "friction" between these properties, in that language, and the given construction; and to prove by the most varied methods or tools (arithmetic induction, but also complex variable functions, for example) if they are "realized" upon this well-ordered, infinite structure. In other words, we need to compare construction principles and proof principles. It is thus like this that we may understand the essential incompleteness of the formal theory of numbers: the (formal) proof principles are weaker than the (conceptual/structural)construction principles, which give us the well-ordering of integer numbers, in this case (see Chapter 2). It should then be clear that this absolutely does not imply that this infinite sequence "preexists" as a conceptorless concept: if five stones were surely already there, at the foot of this mountain, one billion years ago, what was not there was the concept of the number 5, something completely different, nor were the infinitary properties of that number, ordered within the infinite sequence with the others, as for example the solvability of fifth degree equations or the results of many other linguistic/algebraic constructions we know how to make; constructions that are far from being arbitrary, because rooted in a creative mix of significant conceptual methods (logico-formal, results of spatial invariants, regularities, etc.), but not pre-existing human activities. Yet, they are objective, as rooted in invariants and regularities (order, symmetries, ...) that we conceptualize after and by a friction over this world.

Also consider a variant of chess I am inventing right now: a 100 times 100 square, with 400 pieces that have quite varied but not arbitrary rules: very symmetrical finite movements and simulations of natural movements. I then scatter the pieces randomly; what must be demonstrated is that the configuration thus obtained is compatible with (attainable by) the given rules. Can we say that that configuration (a property of the game) was already there, a billion years ago? What is the meaning of that sentence? Worse, I propose a game with an infinity of squares and pieces, ordered with great originality in the three dimensions, but by effective rules (spirals, fractals ...). I call them "spiralu numbers" or "zamburus," and give you infinitary relationships upon these conceptual objects (I describe, using words, infinite subsets, relationships upon this structure or I scatter the pieces randomly). What sense does it make to say that these properties/relationships were already there? That the compatibility of the distributions of the pieces thus obtained were already decided or were valid since ever? Surely, proof will be necessary in order to "verify" it (I prefer: to check if these distributions are "realized" upon the structure, that is to establish friction, by means of proof, between given properties in the language or the geometry of the squares and the game's construction principles). But as long as the infinitary structure, my construction, built in history, a non-arbitrary extension of a practice of squares and of order, is not posed with the rigor of its construction principles, as the locus where to realize, by the friction of proof, this other construction given in the language of the properties to verify, what sense does it make to say that the conceptual structure and the properties of its infinite subsets "pre-existed"? Conversely for the games which I just proposed, which are my own individual construction, the grounding in the world, within a very ancient intersubjectivity, of the concept of the number, of zero, of the successor, of the infinite well-order gives them a "transcendent" status with respect to my individual existence. Yet, this must not lead us to forget that also these mathematical "objects" are concepts, the results of a very structured, phylogenetic and historical conceptual construction, determined by its constitutive hypotheses; they are not a "pre-existing ontology," they do not transcend our human, actually animal existence (as counting is a pre-human activity). The fact that we ignore the totality (what does "totality" mean?) of their "properties" (careful with this word) in no way demonstrates this ontology we so easily confer upon them: we ignore them, just as we ignore the totality of the scatterings of our whimsical chess games on the infinite chessboard above. There is no transcendence in mathematics, or, rather, there is no transcendence which is not the result of non arbitrary constitutive processes (for example, the construction of algebrico-formal enunciations or of the well-ordering of integers), constructions needing to be compared (relatively realized) with one another, by means of this "friction" between and upon conceptual structures, which is called mathematical proof. More specifically, between principles of proof (that we give ourselves, by non-arbitrary choices) and construction principles (that participate in our own cognitive determination, in the relationship with the world).

Continue, for example, and start with the construction of the integers and pass on to the rationals, as ratios of integers, modulo an equivalence of ratios; then consider the convergent sequences (of Cauchy) of these new numbers, modulo equiconvergence. There are the real numbers, constituted using a mathematical method which reconstructs and links together, in its own way, different histories, by distilling the key concepts. The real numbers do not exist, in any sense of a plausible ontology, but their constitution is as objective as are many other conceptual organizations of the world which render it intelligible to us. And they propose to us a very efficient conceptual structure for the phenomenal continuum of time and space. In short, the very existence and the objectivity of Cantor's real numbers is entirely in their construction.

1.2.3 *Laws, structures, and foundations*

In the first part of this chapter, Francis Bailly, from the perspective of physics, poses other important questions, among which I now retain those concerning the terms of "structure" and of "foundation." What I deny is that one can identify the notion of mathematical structure with its axiomatic presentation and, then, that the analysis of proof, within these axiomatic frameworks, can be a sufficient foundational analysis. To discuss this last point, we will also speak of "laws."

Physicists sometimes confuse "formalism" with "mathematization"; it is customary of their language. The mathematical structuration of the world, of a physical experiment, that they propose is often called "formalization." That is quite understandable, because in what concerns the "very concrete" about which they are thinking (physical "reality"), the mathematical structure is surely abstract and symbolic. But with a bit of experience with the debate about the foundations of mathematics, where these terms are employed with rigor (and philosophical relentlessness, I would say), one understands that rigorous, abstract, and symbolic does not mean formal. In fact, a formal system must work without reference to meaning; it is constructed and manipulated thanks only to mechanical rules. These rules are also and surely used during a physico-mathematical calculus, but the formula about which the physicist thinks has nothing to do with that of logical "formalism": the formula is significative from the onset, because the physicist constructed it with permanent reference to its meaning, to his or her physical experience, he or she inserts it into a mathematical context rich with explicative connections. The physicist proposes mathematical structures to make his or her experience intelligible, the physicist does not invent a set of formal rules disconnected from the world, as would do the formalist, whose foundational analysis lies only in consistency. He or she thus proposes mathematical structures, and not formal systems. Between the two there are at least the great theorems of incompleteness, which separate structural construction principles from formal deductions.

Let's try to exemplify this distinction within mathematics themselves. Consider, as "construction principles," translations, and rotations of figures constructed by rule and compass; if one fixes the unit of length, one will easily construct a segment of length the square root of two. And there, a very first challenge for mathematical understanding: the theory of linear equations with integer coefficients, and with its formal rules of calculus, is demonstrably incomplete with regards to this construction (the segment is not a ratio of integers). With the same principles of construction, including the absence of gaps and jumps within the Euclidean continuum, construct the limit of the polygons inscribed in and circumscribed around a circle. It will then be the formal theory of rational coefficient algebraic equations, which is incomplete with regards to this construction of π.

If we move on to the XXth century, Gödel demonstrated that the formal theory of numbers, with its proof principles, is incomplete with regards to the well-order of integers as a construction principle. By analogy to the role of symmetries in physics, one could say in that regard that Hilbert's conjecture of the completeness of formal arithmetic was a mirror-symmetry hypothesis between formal language and ontologizing semantics (the first accurately reflects the second). Gödel's theorem of incompleteness breaks this alleged symmetry and initiates modern logic. In more constructive and recent terms, the breaking of the symmetry between proof principles and construction principles, of an essentially geometric nature, leads us to understand the insufficiency of a sole logico-formal language as the foundation of mathematics and brings back to the center of our forms of knowledge a constitutive mathematics of time and space, thanks to its construction principles. There is concrete incompleteness, a modern version of Gödelian incompleteness, a discrepancy or breaking in provable symmetry between construction principles and proof principles.

Mathematical structures are, in fact, the result of a reconstruction which organizes reality, all the while stemming from concepts, such as the pre-mathematical concept of the infinite (the theological concept, for example), or, even, from pre-conceptual practices (the invariants of memory), the experience and *practice* of order, of comparison, the structurations of the visual and perceptual in general gestalts. These lead to a structuration, explicated in language, of these (pre-)concepts and of their relationships: the well-order of the integers, the Cantorian infinite, the continuum of the real numbers, the notion of a Riemannian manifold. The concept of infinity gets involved, because it is the result of a profound and ancient conceptual practice, as solid as many other mathematical constructions; these practices are not arbitrary and each may be understood and justified by the process of the construction of scientific objectivity to which it is related.

After the construction of these abstract structures that are symbolic yet rich in meaning, because they refer to the underlying practical and conceptual acts of experience, we may continue and establish axiomatic frameworks that we attempt to grasp at a formal level, whose manipulation may disregard meaning. This process is important, because it adds a possible level of generality and especially highlights certain, possible, "proof principles" which enable us to work, upon these structures, by using purely logico-formal deductions, within well specified languages. But these principles are essentially incomplete, that is what the great results of incompleteness of the last 70 years, in particular the recent "concrete" ones, tell us, as we will explain in Chapter 2. Moreover, as we said in the introduction, the analysis of proof, particularly if this analysis is only formal, is but the last part of an epistemology of mathematics: it is also necessary to account for the constitution of the concepts and of the structures which are manipulated during these proofs. But there is more to this usual and fallacious identification of "axioms" with "structures," of "foundations" with "logico-formal rules." In order to understand this, let's return to physics. Husserl, in an extraordinary epistolary exchange with Weyl (see Tonietti, 1988), grasps a central point of relativistic physics, highlighted, particularly, by the mathematical work of Weyl (but also by the reflections of Becker, a philosopher of physics and student of Husserl, see Mancosu and Ryckman, 2002). The passing from classical physics to the new relativistic framework first bases itself upon the following change in perspective: we go from *causal lawfulness* to the structural organization of time and space (*structural lawfulness*), nay, from causal lawfulness to intelligibility by mathematical (geometric) structures. In fact, Riemann is at the base of this revolutionary transformation, all the while developing the ideas of Gauss. In his habilitation memoir, (Riemann, 1854), a pillar of modern mathematics and of their applications to physics, he aims to unify the different physical fields (gravitation and electromagnetism) through the geometrical structure of space. He throws out the hypothesis that the local structure of space (its metric, its curvature) may be "linked to the cohesive forces between bodies." "Divination" Weyl will call it in 1921, for it is effectively the viewpoint peculiar to this geometrization of physics which at least begins with Riemann, finds its physical meaning with Einstein, and, with Weyl, its modern mathematical analysis.

It thus seems to me that the attempt to mathematize the foundational analysis of mathematics by only referring to the "laws of thought" is comparable to a reconstruction of the unique, absolute classical universe in physics, with its Newtonian laws. It is not *a priori* laws that regulate mathematics, but they do constitute themselves as structures, conceptual plays, that are not arbitrary. The "cohesive forces," in mathematics, would correspond to an "interactive dynamic of meaning," a structuration of concepts and of deduction itself.

In category theory, for example, we propose a new conceptual structure, by novel objects (invariants) and morphisms (transformations); we link it to other structures by using functors, that we analyze in terms of transformations ("natural," their technical name), all the while following/reconstructing the open dynamic of mathematics, of which the unity manifests itself through these reciprocal translations of theories (interpretation functors). And the relative (functorial) interpretations relate the ongoing conceptual constructions (categories): unity is an ongoing conquest and not given by a pre-existing set-theoretic background universe.

Moreover, certain of these categories have strong properties of closure, a bit like rational numbers that are closed for multiplication and division, as real numbers are for particular limits. One of the logically interesting properties, among many others, is "small completeness," that is, the closure with regards to products which interpret the universal quantification, among which, in particular, is second-order quantification (quantification upon collection of collections). Through this device, some categories confer mathematical meaning to the challenges of impredicativity (Asperti and Longo, 1991), the great bogeyman of "stratified" worldviews and of logic: the formal certitudes constructed upon elementary and simple building blocks, one level independent of the other. The world, however, seems to build itself upon essential circularities, from the merest dynamical system (three bodies interacting in a gravitational field) or the local/global interaction (non-locality) in quantum physics, up to the "impredicative" unity of any living organism, of which the parts have no meaning and are out of place outside of the organism as a whole. Maybe the emergence of that which is new, in physics, in biology, only takes place under the presence of strong circularities, sorts of internal interactions within complex systems.

Mathematics is thus not a logico-formal deduction, nicely stratified from these axioms of set theory that are as absolute as Newton's universe, but is structurations of the world, abstract and symbolic, doubtless, yet not formal, because significant; its meaning is constructed in a permanent resonance to the very world it helps us understand. They then propose collections of "objects" as conceptual invariants, of which the important thing is the individuation of the transformations which preserve them, exactly like (iso-)morphisms and functors preserve categorical structures (properties of objects of a category).

There are no absolutes given by logical rules, beyond the world and the cognitive subject, by definite rules (but then why not those of scholastics or of Euclid's key rule: "a part always has less elements than does the whole," which is false in the case of our infinite sets?), but there is a dynamic of structures (of categories), emergent from a mathematical practice, then linked by those interpretation functors which unify them, which explain the ones by the others, which confer upon them meaning within a "reflexive equilibrium" of theories (and of categories, particularly those which correspond to deductive systems (Lambek and Scott, 1986; Asperti and Longo, 1991)).

Surely there is a temporality in the construction of the meaning we confer to the world through mathematics; and it is a "rich" temporality, because it is not that of sequential deduction, of Turing machines: it is closer to the evolution of space-distributed dynamical-type systems, as we shall see. We must let go of this myth of pre-existing "laws of thought" and immerse mathematics into the world while appreciating its constitutive dynamics of which the analysis is an integral part of the foundational project. The laws or "rules" of mathematical deduction, which are surely at the center of proof, are themselves also the constituted of a praxis, of language, as invariants of the reasoning and of the practice of proof itself.

The foundation, so, as the constitutive process of a piece of knowledge, is constructed responsively to the world, the physical world and that of our sensations. But ... where does this process begin? It is surely not a case of reascending "to the mere stuff of perception, as many positivists assert," since physical objects are "intentional objects of acts of consciousness" (Weyl, 1918a). There is a very Husserlian remark, a constitution of objects which we have called a conceptual "decoupage" (cutting-off). And this deecoupage is performed (and produced) by the mathematical concept, a conscious (intentional) act towards the world. Then, reasoning, sometimes rooted in a whole different practice, in the language of social interaction, that of the rules of logical coherence or of the aesthetic of symmetries, for example, generates new mathematical concepts, which may themselves, but not necessarily, propose new physical objects (positrons, for example, derived from electrons by a pure symmetry in equations in microphysics).

The autonomy of mathematics, thanks to the generativity of reasoning, even of the formal type (calculus for example), is indubitable, there lies its predictive force in physics. The integration of these different conceptual dimensions, of these different praxes (geometrical structuration of the world, logical and formal deduction, even far removed from any physical meaning), also confers upon mathematics its explicative and normative character with regards to reality: one goes, in space, let's say, from a physical invariant to another by purely logico-formal means (an algebraic transformation applied to this invariant and which preserves it, a symmetry ...) and a new physical object is thus proposed. The physical proof will be a new experience to invent, with instruments to be invented.

Obviously, in this grounding of our sciences in the world, perception also plays an essential role, but we must then develop a solid theory of perception, rooted in a cognitive science that allows us to go far beyond the positivist's "passive perception," of which Weyl speaks about. We shall return to this point.

The approach we propose, of course, causes the loss of the absolute certitude of logico-formal, decidable proof. But we know, since Gödel, that any formal theory, be it slightly ambitious and of which the notion of proof is decidable, is essentially incomplete. So logicism's and formalism's "unshakeable certainties" (the absolute certification of proof) are lost ... since a long time. There remains the risk of the construction of scientific objectivity, thoroughly human, even in mathematics, the adventure of thought which constitutes its own structures of the intelligibility of the world, by the interaction with the former and with the thoughts of others. The risk we will take in Chapter 2 of acknowledging the foundational role of the well-ordering of integers, by a geometric judgment constituted in history, action, language, and intersubjectivity, will be to certify the coherence of arithmetic.

1.2.4 Subject and objectivity

In various works, Weyl develops a very interesting philosophical analysis concerning the passage in physics from the subjective to the objective, on the basis of references to his own mathematical works in relativity theory. This analysis is emphasized by Mancosu and Ryckman (2002), who refer mostly to Weyl (1918a-b, 1927). The importance of Weyl's remarks obviously extends way beyond the philosophical stakes in physics and in mathematics, because it touches upon a central aspect of any philosophy of knowledge, the tension between the "cult of the absolute" and "relativism." Husserl seeks to move beyond this split in all of his work and in his reading of the history of philosophy (see, for example, Husserl, 1956). XXth century physics can provide tools for contributing to that debate, and those are Weyl's motivations.

For Weyl, immediate experience is "subjective and absolute," or, better, it claims to be absolute; the objective world, conversely, that the natural sciences "crystallise out of our practical lives ... this objective world is necessary relative." So, it is the immediate subjective experience which proposes absolutes, while the scientific effort towards objectivity is relativizing, because "it is only presentable in a determined manner (through numbers or other symbols) after a coordinate system is arbitrarily introduced in the world. This oppositional pair: subjective-absolute and objective-relative seems to me to contain one of the most fundamental epistemological insights that can be extracted from natural sciences".

Following his works in relativity, Weyl thus gives a central role to reference frames. The subject lays, chooses, a reference frame and in this manner organizes time and space. That choice is the very first step performed by the knowing subject. But the operation of measurement, by means of its own definition, also implies the subject: any physical size is relative to (and set by) a "cognizing ego." The passage to objectivity is given, in quantum physics, by the analysis of "gauge invariants," for example, one of Weyl's great mathematical contributions to this field: they are given as invariants in relation to the passing from one reference and measurement system to another. More generally, the passage from subjectivity to scientific objectivity implies the explicit and explicated choice of a reference frame, including for mathematical measurements. The analysis of the invariants with respect to different reference frames gives then the constituted and objective knowledge.

Weyl thus emphasizes, in Husserlian fashion, that any object in the physical world is the result of an intentional act, of the awareness "of a pure, sense giving ego." For both thinkers, it is a matter of the Cartesian, "Ego" to which Husserl so often returns to, which "is, since it thinks"; and it is, because, as a consciousness, it has "objects of consciousness" (consciousness is "intentional," it has an "aim"). It is the subject, this conscious Cartesian "Ego," that chooses the reference frame and who, afterwards, is set aside. It poses the origin, the 0 and the measurement, and it mathematically structures time and space (as a Cantor-Dedekind continuum, for example, or as a Riemannian manifold with its curvature tensors); by that act (the construction of a space as a mathematical manifold), it poses a framework of objectivity, independently of the subject, objectivity nevertheless consciously relativized to that choice. Because the choice of viewpoint, of the frame, is relativizing and breaks the absolute characteristic of the subject before the passage to scientific objectivity; this passing of subjectivity, which claims to be absolute, to relativizing objectivity, is the meaning of the scientific approach central to relativity. Just as it is very well put in (Mancosu and Ryckman, 2002): "The significance of [Weyl's] 'problem of relativity' is that objectivity in physics, that is, the purely symbolic world of the tensor field of relativistic physics, is constituted or constructed via subjectivity, neither postulated nor inferred as mind-independent or transcendent to consciousness." But this symbolic world of mathematics is in turn itself the result of an interaction of the knowing subject(s), within intersubjectivity, with the regularities of the world, these regularities, which we see and which are the object of intentional acts, of a view directed with "fullness and willing," as Husserl and Weyl say.[6]

The subject is thus at the origin of scientific knowledge, and it is with the subject that any mathematical construction begins. However, it will be necessary to push the analysis of the subject's role further: today we can pose the problem of objectivity at the very center of the knowing subject, because this subject is not the psychological subject, which is also disputed by the seekers of the absolute, of transcendental truths, of configurations or properties which are already there, true prior to any construction/specification, even in my infinite chessboard or in the sequence of integers. In fact, it is a question of the "cognitive subject," of this "Ego" that we share as living, biological creatures, living in a common history that is co-constituted with the world, at the same time as its activity in the world. There is the next issue we will have to deal with, in the dialog with cognitive sciences, basing ourselves on non-naive (and non passive) theories of perception, on theories of the objective co-constitution of the subject. The scientific analysis of the subject must, by these means, underline what is common to subjective, psychological variability: more than a simple "intersection of subjectivities" it is a question of grasping in that way what

---

[6]The profoundness of Weyl's philosophy of sciences is extraordinary and his philosophy of mathematics is but a part of it (a small one). I found quite misleading, with regards to this profoundness and its originality, the many attempts of many, including some leading "predicativists" to make him into a predecessor of their formalist philosophy of mathematics. Briefly, in Weyl (1918a), a remarkable Husserlian analysis of the phenomenal continuum of time and space, Weyl also feels concerned with the problem of "good definitions," a problem that preoccupied all mathematicians of the time (including Poincaré and Hilbert, of course): the XIXth century was a great period for mathematics, but, so very often, ...what a confusion, what a lack of rigor! Particularly, it was necessary to watch out for definitions that may have implied circularities, such as impredicative definitions. Weyl noticed that Russell's attempt to give mathematics a framework of "stratified" certitudes does not work ("he performs a hara-kiri with the axiom of reductiveness," (Weyl, 1918a)). In the manner of the great mathematician he was, Weyl proposed, in a few pages, a formally "predicativist" approach that works a thousand times better than that of Russell with his theory of types. An approach and an interesting exercise in clarification that Weyl will never follow in his mathematical practice; to the contrary (Weyl, 1918a), he criticizes, quite a few times, Hilbertian formalism of which the myth of complete formalization "trivializes mathematics" and comes to conjecture the incompleteness of formal arithmetic (!). Feferman (1987) took up these ideas only slightly brushed upon by Weyl (and not the predicatively incoherent heaviness of Russell's theory of types), to make it into an elegant and coherent predicative formal theory for analysis. Remarkable technical work, but accompanied by an abusive and quite incomplete reading of Weyl's philosophy, which is a much broader philosophy of natural sciences, never reduced to stratified predicativisms nor their corresponding formal or logicist perspectives.

lies behind individual variabilities, what directs them and allows them to communicate and to understand/construct the world together.

Foundational analysis, in mathematics and in physics, must therefore propose a scientific analysis of the cognitive subject and, then, highlight the objectivity of the construction of knowledge within its referential systems or reference frames.

In what concerns the foundations of mathematics, a process analogous to this "choice of reference frame" is well explicated, in category theory, by choosing the right "topos" (as referential category for a logic or with an "internal logic" (Johnstone, 1977)), to relate, through interpretation functors, other categorical constructions, in a dynamic of these structures by which we give mathematical meaning to the world (algebraic, geometrical, manifolds' categories, ...). This has nothing to do, as we have already emphasized, with the absoluteness of the axioms of set theory, a Newtonian universe that has dominated mathematical logic and that has contributed for a century to the separation of mathematical foundations from epistemology and from the philosophy of natural sciences. That was a matter, indeed, of an absolute, that of sets, intuition of which is compared, as we said, by the "realists" in mathematical philosophy, to the perception of physical objects (quite naively described in its passivity), sets and objects also being transcendent, with their properties all listed there, "pre-existing." A typical example of that which Husserl, de Ideen, and Weyl (taken up by Becker, see Mancosu and Ryckman, 2002) call the "dogmatism" of those who speak of absolute reality, an infinite list of already constituted properties, constituted before any pre-conscious and conscious access, before projecting our regularities, interpreting, acting on the world, before the shared practices in our communicating community.

1.2.5 *From intuitionism to a renewed constructivism*

Quite fortunately, within the same mathematical logic, we begin to hear different voices: "Realism: No doubt that there is reality, whatever this means. But realism is more than the recognition of reality, it is a simple-minded explanation of the world, seen as made out of solid bricks. Realists believe in determinism, absoluteness of time, refuse quantum mechanics: a realist cannot imagine 'the secret darkness of milk.' In logic, realists think that syntax refers to some pre-existing semantics. Indeed, there is only one thing which definitely cannot be real: reality itself" (Girard, 2001). The influence of Brouwer, the leader of intuitionism, and of Kreisel, as well as the mathematical experience with intuitionist systems, is surely present in the mathematical work and in the rare philosophical reflections of Girard, but without the slip, characteristic of Brouwer, into a senseless solipsism, nor with the *a priori* limitations of our proof tools. Moreover, time and space are included in Girard's proof analysis: the connectivity, the symmetries of proof as network, time as irreversible change in polarity in Girard (2001), have nothing to do with "time as secreted by clocks" (his expression), the time of sequential proof, of Turing machines, which is beyond the world (see Chapter 5).

Brouwer's intuitionism, among the different trends in the philosophy of mathematics (formalist, Platonic realist, intuitionist), is possibly the only foundational analysis that has attempted to propose an epistemology of mathematics (and a role for the knowing subject). The discrete sequence of numbers, as a trace of the passing of time in memory ((Brouwer, 1948), see also (Longo, 2002a, 2005)), is posed as constitutive element of mathematics. It is exactly this vision of mathematics as conceptual construction that has made Weyl appreciate Brouwer's approach for a long time. In fact, the analyses of the mathematical continuum for Brouwer and Weyl (as well as for Husserl, see (Weyl, 1918a; Tonietti, 1988; Longo, 1999)) are quite similar in many respects. However, Weyl had to distance himself from Brouwer, during the 1920s, when he realized that the latter excessively limits the tools of proof in mathematics and does not know how to go beyond the "psychological subject," to the point of renouncing the constitutive role of language and of intersubjectivity and to propose a "languageless mathematics" (a central theme of Brouwer's solipsism, see (Brouwer, 1948; van Dalen, 1991).

Conversely, and as we have tried to see, the relativity problem for Weyl, as a passage from "causal lawfulness" to "structural lawfulness" in physics, as well as a play between subjectivity-absolute and objectivity-relative, is at the center of an approach that poses the problem of knowledge in its unity, particularly as it is the relationship between physical objectivity and the mathematical structures that make time and space intelligible, thanks, among other things, to language. All the while following Weyl, we have made a first step towards an extension of foundational analysis in mathematics by a cognitive analysis of what should precede purely logical analyses: only the last segment is without doubt constituted by the logico-formal analysis of proof. But upstream there remains the problem of the constitution of structures and of concepts, a problem which is strictly related to the structuration of the physical world and to its objectivity. The project of a cognitive analysis of the foundations of mathematics thus requires an explication of the cognitive subject. As a living brain/body unit, dwelling in intersubjectivity and in history, this subject outlines the objects and the structures, the spaces and the concepts common to mathematics and to physics on the phenomenal veil, while constituting itself. In short, parallel constitutive history, in physics, begins with perception as action: we construct an object by an active viewing, by the presence of all of our body and of our brain, as integrator of the plurality of sensations and actions, as there is no perception without action: starting with Merleau-Ponty's "vision as palpation by sight," perception is the result of a comparison between sensorial input and a hypothesis performed by the brain (Berthoz, 1997). In fact, any invariant is an invariant in relation to one or more transformations, so in relation to action. And we isolate, we "single out," invariants from the praxis that language, the exchange with others, forces us to transform into concepts, independently, as communicables, from the constitutive subject, from invariants constituted with others, with those who differ from us but who share the same world with us, and the same type of body. From the act of counting, the appreciation of the dimensionless trajectory – dimensionless since it is a pure direction – we arrive at the mathematical concepts of number, of aunidimensional line, with no thickness, and, then, of a point. Invariants quite analogous to the physical concepts of energy, force, gravitation, electron. The latter are the result of a similar process, they are conceptual invariants which result from a very rich and "objective" praxis, that of physics, inconceivable without a close interaction with mathematics. They organize the cues that we select through perception and through action upon the world, through our measurement instruments; the geometrical structuration of those invariants is the key organizing tool, because it explicates in time and space our action and our comprehension. These are the cognitive origins of the common construction principles in mathematics and physics.

Individual and collective memory is an essential component to this process constitutive of the conceptual invariants (spatial, logical, temporal, ...). The capacity to forget in particular, which is central to human (and animal) memory, helps us erase the "useless" details; useless with regards to intentionality, to a conscious or unconscious aim. The capacity to forget thus contributes in that way to the constitution of that which is stable, of that which matters to our goals, which we share: in short, to the determination of these invariant structures and concepts, which are invariant because filtered of all which may be outside our intentional acts of knowledge. Their intercultural universality is the result of a shared or "sharable" praxis, in the sense that these invariants, these concepts, may very well be proposed in one specific culture (think about Greek geometry or Arabic algebra), but their rooting in fundamental human cognitive processes (our relationship to measurement and to the space of the senses, basic counting and ordering, ...) make them accessible to other cultures. This widening of a historic basis of usage is not neutral, it may require the blotting out of other experiences specific to the culture which assimilates them, but confers on them this universality that accompanies and which results from the maximal stability and conceptual invariance specific to mathematics. But this universal is posed with relation to human forms of life and does not mean absolute; it is itself a cultural invariant, between cultures that take shape through interaction. Because universality is the result a common evolutive history as well as of these communicating communities. Historical demise is a factor of it: oblivion or expulsion from mathematics of magical numbers, of "zombalo" (whatever) structures ... of that which does not have the generality of method and results we call, *a posteriori*, mathematical.

As for the mathematical organization of space, both physical and sensible, it begins very early, probably as soon as space is described by gesticulation and words, or with the spatial perspective and width of the pictorial images of Lascaux, 20,000 years ago, or from the onset of the play of Euclid's rigid bodies, which structures geometrical space. Euclid's axiomatics indeed summarize the minimal actions, indispensable to geometry, with their rule and compass, as construction and measurement instruments: "trace a straight line from one point to another," "extend a finite line to a continuous line," "construct a circle from a point and a distance" ... (note that all these constructions are based and/or preserve symmetries). His first theorem is the "vision of a construction" (in Greek, theorem means "sight," it has the same root as "theater"): he instructs how to "construct an equilateral triangle from a segment," by symmetric tracing with a compass.[7]

---

[7] That's what the first theorem of the first book is and the point constructed during the proof is the result of the intersection of two lines traced using a compass. Its existence has not been forgotten, as claimed by the formalist reading of this theorem, it is constructed: Euclid's geometry presupposes and embraces a theory of the continuum, a cohesive entity without gaps or jumps (see Parmenides and Aristotle). Note that the concept of the dimensionless point is a consequence of the extraordinary Greek invention of the line with no thickness: points (semeia, to be precise – signs) are at the extreme of a segment or are obtained by the intersection of two lines, a remark by Wittgenstein.

This history leads to Weyl's symmetries, regularities of the world which impose themselves (donations that, in this sense, pre-exist or that reality imposes on us), but that we see or decide to see. We then transform them into concepts and choose to pose them as organizing criteria of reality, even in microphysics, far removed from sensorial space.

1.3 Regarding Mathematical Concepts and Physical Objects

(by Francis Bailly)

Giuseppe Longo suggests that we establish a parallel between *mathematical concept* and *physical object*. The massive mathematization of physics, the source of new mathematical structures it is thus likely to generate, the aptitude of mathematics to base its constructions on the physicality of the world, possibly in the view of later surpassing or even of discarding this physicality in its movement of abstraction and its generativity proper, all these entail questions regarding the relationships that the mathematical concepts thus constructed may entertain with the highly formalized physical objects of classical or contemporary physics. So what could be the static or dynamic traits specific to each of these determinations and of these methods which would permit such a parallel?

In the introduction, we have already stated the analogies and differences between the foundations of physics and the foundations of mathematics in the respective corresponding of their construction principles and of their proof principles. In short, they seem to share similar construction principles and to have recourse to different proof principles – formal for mathematics *via* logic, and experimental or observational for physics *via* measurement. Is it possible to go further without limiting ourselves to the simple observation of their reciprocal transferals?[8]

---

[8] There is no need to return to the obvious and constitutive transfer from mathematical structures to physics. The transfer in the other direction, from physics towards mathematics, is on the other hand more sensitive. Recall the introduction in physics of the Dirac "function," which led to distribution theory; or the introduction of Feynman path integrals and the corresponding mathematical research aiming to provide them with a rigorous foundation; more recently, the Heisenberg non-commutative algebra of quantum measurement and Connes' invention of non-commutative geometry. Without speaking of the convergences between the physical theory of quasi-crystals and combinatory theory in mathematics or between physical turbulence theory and the mathematical theory of non-linear dynamical systems, etc.

1.3.1 *"Friction" and the determination of physical objects*

To address a different level, let's note that in the epistemological discussion regarding the relationships between the foundations of physics and the foundations of mathematics, Giuseppe Longo proposes that we consider that which causes *"friction"* in the set of determinations of their respective objects. For physical friction, that which validates it (or serves as proof), can be found first of all in the relationship to physical phenomenality and its measurement: experience or observation are determinant in the last instance, even if at the same time more abstract frictions continue to operate, frictions which are more "cognitive" with regard to mathematical theorization. In contrast, for mathematics, it very well appears that the dominant friction may be found in the relationship to our cognitive capacities as such (in terms of coherence of proof, of exactitude of calculation), even if mathematical intuition sometimes feeds on the friction with physical phenomenality and may also be canalized by it.

Let's explain. "Canalization" and "friction," in the constitution of the mathematical concepts and structures of which Longo speaks, seem mainly related to *"a reality"* resulting from the play between the knowing subject and the world, a play which imposes certain unorganized regularities. Mathematical construction then and again enters into friction with the world, by its organization of reality. In physics, where prevails the *"blinding proximity of reality"* (Bitbol, 2000a), friction and canalization seem to operate within distinct fields: if friction, as we have just highlighted, remains related to the conditions of experience, of observation, of measurement – in short, of physical phenomenality – canalization, on its part, now results much more from the nature and the generativity of the mathematical structures which organize this phenomenality, modelize it and finally enable us to constitute it into an objectivity. In a provocative, but sound, mood many physicists consider the electron to be just the solution of Dirac's equation.

If we refer to the aphorism according to which *"reality is that which resists,"* it appears that, by interposed friction, physical reality, all the while constituting itself now *via* mathematization, finds its last instance in the activity of the measurement, while mathematical phenomenality may be found essentially in the activity associated with our own cognitive processes and to our abstract imagination. What relationship is there between each of these types of friction, between both of these realities? At a first glance, it does appear that there is none: the reality of the physical world seems totally removed from that of the cognitive world and we will not have recourse to the easy solution which consists in arguing that in one case as in the other, we have to face material supports, relative to a unique substantiality. Indeed, even if it is the same matter, it manifests at very different levels of organization which are far removed the one from each other according to whether we are considering physical phenomenality or our cognitive structures. On the other hand, if we are nevertheless guided by a somewhat monist vision of our investigational capacities, we can legitimately wonder about the coupling between these levels enabling knowledge to constitute itself: coupling dominated by one of the poles involved (phenomenal or cognitive) according to whether we use a physical or a mathematical approach. In fact, it is this coupling itself which appears to constitute knowledge, as well as life itself, if we agree that cognition begins with life (Varela, 1989). The relationship that could then be established between the physical object and the mathematical concept would therefore stem from the fact that it would be a question of accounting for the same coupling (between cognitive structures and physical phenomenality) but taken from different angles depending on whether it is a question of a theory of physical matter or a theory of abstract structures. The fact that it is the same coupling would then manifest within the community of construction principles which we have already described, whereas the difference in disciplinary viewpoints regarding this coupling (from the phenomenal to the cognitive-structural or conversely) would manifest in the difference in nature of the proof principles, demonstration, or measurement.

Another approach for conducting a comparison between physical object and mathematical concept, which seems to ignore the phenomenal friction characteristic of physics, consists in considering that the scientific concept, be it attached to a physical object or to a mathematical ideality, has lost a number of its determinations of "concept" to become an abstract formal structure. This would only translate the increasing apparentness between contemporary physical objects and mathematical structures modelizing them. It could possibly be a way to thematize the constitutive role of mathematics for physics, which we have already presented. Such an appreciation is probably well founded, but it does not completely do justice to this particular determination of physical objects to be found in the second part of the expression "the mathematical structures *which modelize them.*" Indeed, the necessity of adding this precision refers to this "something" which needs to be modelized, to a referent which is not the mathematical structure itself. And it is this change of level of determination, this heteronomy contrasting with the autonomy of the mathematical structure, which probably specifies the objectivity of the physical object and suggests that it is likely to respond to other types of determination than those attached to the sole mathematical structure, to that which we can henceforth call a "new" friction.

1.3.2 *The absolute and the relative in mathematics and in physics*

The physicist can only fully agree with what Weyl had indicated when producing a critique, from the mathematical standpoint, of the so-called absoluteness of the subjective in order to confront it with what he would call the relativity of objectivity (see the text by Longo above). All of his work consists indeed in breaking away from the illusion of this subjective "absoluteness" in the apprehension of phenomena in order to achieve the construction of objective invariants likely to be communicated. By doing this, he obviously succeeds in qualifying the subjective as relative and in qualifying the objective if not as absolute, at least as stable invariant. Moreover, the results of this construction of physical objectivity prove to be sometimes incredibly counter-intuitive. Consider for example quantum non-separability which prohibits speaking of two distinct quanta once they have interacted; but already, in the age of Copernicus and Galileo, the roundness of the world and its movement relative to the sun was a challenge to intuitive spontaneous perception and common sense: language perpetuates traces of this, it continues to see the sun rise.

What is at stake in the analysis one can make of this situation, can doubtlessly be found in the relationships between the use of natural language on the one hand and the mathematization presiding over the elaboration of mathematical models on the other hand. To put it briefly, the relativity of the subjective is relative to language and may thus appear to be absolute since language then plays a referential role, whereas the relativity of the objective is relative to the model itself that is presented as a source of the stable invariants which, by this fact, are likely to play a role of "absolutes" in that they are (in principle) completely communicable. Hence the possibility for a sort of chiasma in the qualifications according to the first referral, meaning also according to the involvement of the person speaking: in his or her intuitive and singular grasp only the absoluteness of the intuition is perceived and not the relativity of the "ego." But in the rational reconstruction aiming for objectivity, relativity is not included within the mathematical model and the invariant objectives constructed by the intersubjective community are taken for quasi "ontological" absolutes. It is these referrals (to language and to the mathematical model) which we will therefore attempt to discuss more precisely.

1.3.3 On the two functions of language within the process of objectification and the construction of mathematical models in physics

Despite the observation of the increasingly mathematical and abstract character of the scientific object of physics, it would nevertheless be wrong to conclude that the corresponding scientification movement only comprises a process of removal with regard to natural language and a discredit of its usage. This would be to completely ignore that scientific intuitions, however formal they may be, continue to be rooted in this linguistic usage and that the interpretations which contribute to making them intelligible, including for the constitutive intersubjectivity of the scientific community, cannot do without these intuitions. The references to ordinary language and common intuitions are needed in order to communicate not only with the non-specialists of the field, but also in the heuristic of the disciplinary research itself, notably in its imaginative and creative moment, as singular as it may be.

Thus, under deeper analysis, objectifying mathematization appears in fact not to be exclusive to ordinary language and its usage, even if in the moment of the technical deployment it may be as such. It rather appears to be an insert between two distinct functions of language, which it contributes to distinguish all the while articulating them. By doing this, the hermeneutical dimension is reactivated and the components of history and genesis are reintroduced also to where mainly dominates the mathematical structures and their conceptual and theoretical organizations. It is this point we would like to argue and develop here by asserting that in the same way that we are led to consider, for reason, a double status – constituting and constituted reason – we are led to distinguish two functions for language, relatively to mathematical formalism: a *referring* function and a *referred* function. Let's try to clarify this.

In its referring function, language provides the means of formulating and establishing, for physics (but this also applies to other disciplines), the major theoretical principles around which it organizes itself. Relatively to the norm-setting subject, in a certain sense it governs objectifying activity. In contrast, in its referred function relatively to these modelizations, language tends more to use *terms* (mathematical terms, typically) than words.[9] Similarly, it uses conceptual or even formal relationships rather than references to signification. It is then submitted to the determinations specific to these abstract mathematical structures that it had contributed in implementing and of which it triggered the specific generativity. This is done until the movement of scientific theorization uses this referred state of language to confer it with a new referring function in view of the elaboration of new models, of new principles, that are more general or more abstract, the "final state" of a stage becoming in a way the "initial state" of the following stage. In this always active dialectic process, mathematics maintains the gap and the distinction – which are essential for the construction of objectivity – between these two functions of language, all the while ensuring the necessary mediation between them. Its role is reinforced and modified by its informal use in a referring function, while it continuously transforms the referred function by means of the internal dynamic specific to it, due to the generativity of the mathematical model. In doing so, mathematics contributes in generating the language of knowledge by means of the functions it confers to language, between which it ensures the regulated circulation (though in the realm of objectifying rigor, a bit like what poetry achieves in the realm of the involvement of subjectivity).

A physical example of this process, in direct continuity with the innovations by Kepler, Copernicus, and Galileo, may be found in the status of Newton's universal gravitational theory. The referring state of language had recourse in the past to an "Aristotelian" representation of the world according to which the "supralunar" constituted an *absolute* of perfection and of permanence (invariability of the course of planets describing perfect circles, and the corresponding mathematical model of Ptolemaic epicycles). It was, however, from within its referring function (of which the still quasi-mythical state can also be found in Newton's alchemical or Biblical works (Verlet, 1993)) that the mathematical model of universal gravitation was constructed, as Galileo had wanted. This model governs all bodies, be they infra- or supralunar. Thanks to mathematization, this radical relativization as for interaction forces, is indeed accompanied by the maintaining (or even, by the introduction) of another *absolute*, that of time and space. However, it also redefines the language of the course of planets in a state that is

---

[9]In the sense that it often uses, as terms, more or less precise, more or less polysemic, words from ordinary language to designate much more rigorous (of no or restricted polysemy) and abstract notions which have meaning only within the "technical" context within which they are used.

henceforth referred to in this model where elliptic orbits and empirical observations are "explained" by the law of universal gravitation. Even more: the relevant physical invariants that will serve as support structure for any ulterior consideration and which will model the language of this new cosmology are identified. It is this referred state of the language of Newtonian cosmology (the mathematical model thus constructed) that will afterwards serve as the new foundation for following research, having from that moment a referring role. The geometrization of the Newtonian language of forces will lead to the relativization of the absolutes of space and of time themselves. This will be obtained by its referring use, as the background language for conceiving the Einsteinian theory of general relativity.

Let's return to this distinction from a complementary point of view, which is closer to the procedures, closer also to more specifically logical formalisms. As referred, language must in one way or another, in order to make sense and to avert paradoxes, conform to a sort of theory of types capable of discrimination between the different levels of its statements: it must clearly distinguish between terms and words as well as between different *types* of terms. But the construction of such a theory of types has recourse to the referring function of language. This function guides the conceptual elaboration and the formulation of formal statements. Hence, the referring function invents and is normative, beyond the normative creativity of the referred mathematical model. It governs creative and organizing *activity*. Since the referred function is the object of study and of analysis, it requires the mediation of a logical-mathematical language which objectifies and enables us to process it with rigor also from the point of view of its own referring function.

Let's note at this stage that the requirement of an "effective logic," sometimes formulated by constructivist logicians, concerns essentially the referred role of language. With intuitionism, the activity of thought, beyond language, is at work in the process of mathematical elaboration and construction. While it innovates and creates, that is, as it brings forth new referrals, the activity of thinking does not respond to criteria or norms of constructability: it creates them.

We agree on this creative role, but, by allowing a constitutive and double function to language, our analysis of knowledge construction in science describes the modelization of (physical) theory itself as derived but nevertheless as determinant. While the reflection, as abstract and rigorous as it may already be and of which the anteriority may confer it with a status of apparent absoluteness, enunciates the principles – what is to be modelized

– the mathematical model governs any ulterior theoretical advance. Moreover, these advances may go as far as contradicting principles considered beforehand as evident.

It is this conceptual configuration which enables us to understand how such counter-intuitive situations addressed by contemporary physical theories (quantum mechanics, typically) may nevertheless be "spoken" in a natural language which continues to spontaneously claim the opposite of the results obtained. This so-called natural language is no longer as natural as that: terms have been substituted for its words. And in any case it is no longer based on its specific linguistic structures and grammar (and the corresponding mentalities) but rather on the mathematics of the model which it interprets and comments. However, this use of natural language does not restitute their profoundness and, most important, their generativity which is only proper to the mathematical model, but nevertheless it continues to ensure cultural communication.

1.3.4 *From the relativity to reference universes to that of these universes themselves as generators of physical invariances*

What Weyl highlights is a process of conceptual emancipation: moving away from the "absolutist" illusion of the subjective and of the language-related, scientific concepts are objectified *via* their relativization to the reference universes and their mathematization. But contemporary physics doubtlessly enables a supplementary passing, a transition in this process of emancipation relative to "sensible" constraints, this time through gauge theories of which the very same Weyl was one of the proponents. Indeed, these theories amount to relativizing reference universes by stripping them of most of their properties which were previously conceived as "absolute": space is such that there is no assignable origin for translations or for rotations, resulting in the invariances of the kinetic and rotational moments; time is such that there is no privileged point enabling an absolute measurement, resulting in the invariance of energy in conservational systems. This is for external reference universes. But for internal quantum universes, it goes as follows: the global absence of an assignable origin for the phases of an electronic wave function will entail the conservation of the electric charge; its local absence is the source of the electromagnetic field. Hence, the same fields of interaction are associated with gauge changes authorized by these relativizations of the reference universes. And these intrinsic relativizations are none other than the symmetries presented by these universes (connexity, isotropy, homogeneity, ...). To summarize, *the less, by these symmetries, we are able to specify these reference universes in an absolute way, the more the physical invariances (and abstract determinations) we are able to bring forth.* Add to this the complementary aspect in which the particular specifications of the objects in question are increasingly referred to spontaneous breaking of symmetry. It is as if, beyond the construction of objectivity itself, it was the *very identity* of the object so constructed – in its stability and in its specific "properties" – which was being determined.

1.3.5 *Physical causality and mathematical symmetry*

Longo also notes that with relativistic physics, according to Weyl, among others, there occurs a *"change of perspective: we pass from 'causal laws' to the structural organization of space and time, or even from causal laws to the 'legality/normativity' of geometric structures."*

One can only observe that this movement has since only reasserted and amplified itself, while producing new and difficult epistemological questions. Indeed, we have witnessed, with quantum physics, an apparently paradoxical development: on the one hand, geometric concepts have become omnipresent (be it an issue of topology, of algebraic geometry or, most of all, of symmetries) at the same time that, on the other hand and as we have abundantly emphasized in the "physical" section of this text concerning space and time, quantum events find their most adequate description in unexpected spaces, increasingly removed from our intuitive space-time, or even from relativistic space-time (functional Hilbert spaces, Fock spaces, etc.). Moreover, as we know, the very notion of trajectory is problematic in quantum physics and that causality *stricto sensu* (that which may be associated with relativistic theories) finds itself to be profoundly thrown into question, particularly in the process of measurement; hence, the difficulties in terms of unification between general relativity and quantum theory.

It is therefore necessary to be clear on this: the massive geometrization of quantum physics does in fact amount to having recourse to and working with concepts of a geometrical origin, but the geometry in question is increasingly removed from that of our habitual space-times, be they four-dimensional. In fact, this geometrization is much more associated with the use of symmetries and symmetry breaking, which enable us to both identify invariants and conserved quantities and to mathematically and conceptually construct gauge theories, as we have seen, disjoining yet attempting to articulate internal space-times and external space-times. From this point, we indeed witness the aforementioned observation of "causal laws" by these "mathematical structural organizations," but which apply this time to spaces and times presenting new characteristics.

As noted by Chevalley in his *Presentation* in the work by van Fraassen (1989), this observation is massive, to an extent, in the analysis of contemporary physics and in its very functioning, of " *substituting to the concept of law that of symmetry*," thus extending the appreciation of the author who, while adopting what he calls the "semantic approach" in the analysis of physics, does not hesitate to assert regarding symmetry that: "*I consider this concept as being the principal means of access to the world we create in theories.*"

Such radicalism could surprise at first sight but may, however, be nicely explained if one realizes that it is an issue of considering essential elements of the very process of the construction of physical objectivity and of the determinations of the corresponding scientific objects. Indeed, as conservations of physical quantities are associated with the principles of relativity and of symmetry and that all of physics bases itself on the measurement of quantities related to properties which must remain stable in order to be observed, one may go as far as asserting that these relativities and symmetries, while appearing to reduce the possible information relative to the systems under study, are constitutive of the very *identity* of these systems. As if it were a question, to use the old vocabulary of medieval scholastics, of identifying the primary qualities (that is, their essential identity structures) while leaving with the "laws" the care of regulating their secondary qualities (to which would correspond here their actual behavior). To continue a moment on this path, we could even go so far in the analysis of the relationships between symmetry and identity by considering that all information is a (relative) breaking of symmetry and that, reciprocally, any relative breaking of symmetry constitutes an objective element of information. According to such a schema, it would then be relevant to consider that to the metaphysical substance/form pair, to which was partially substituted during the scientific era the energy/information (or entropy) pair, finally corresponds the symmetry/breaking of symmetry pair as constituent of the identity of the scientific object, which we have just mentioned.

Let's specify our argument concerning the relationships between causal laws and mathematical structures of geometry. Relativistic theories – general relativity in particular – constitute the privileged domain where was first identified, in modern terms, the relevance of causal laws by their identification to structural organizations. This stems essentially from the intrinsic duality existing between the characterization of the geometry of the universe and that of energy-momentum within that universe. By this duality and the putting into effect of the principle of invariance under the differentiable transformations of space-time, the "forces" are relativized to the nature of this geometry: they will even appear or disappear according to the geometric nature of the universe chosen *a priori* to describe physical behaviors. Now, it is similar for quantum physics, in gauge theories. Here, gauge groups operate upon internal variables, such as in the case of relativity, where the choice of local gauges and their changes enable us to define, or conversely, to make disappear, the interactions characterizing the reciprocal effects of fields upon one another. For example, it is the choice of the Lorentz gauge which enables us to produce the potential for electromagnetic interactions as correlates to gauge invariances.

Consequently, if one considers that one of the modalities of expression and observation of the causal processes is to be found in the precise characterization of the forces and fields "causing" the phenomena observed, then it is apparent that this modality is profoundly thrown into question by the effects of these transformations. Not that the causal structure itself will as a result be intrinsically subverted, but the description of its effects is profoundly relativized. This type of observation therefore leads to having a more elaborate representation of causality than that resulting from the first intuition stemming from classical behaviors. Particularly, the causality of contemporary physics seems much more associated with the manifestation of a formal solidarity of the phenomena between themselves, as well as between the phenomena and the referential frameworks chosen to describe them, than to an object's "action" oriented towards another in inert space-time, as classical mechanics could have accredited the idea. After Kant, it already had ceased claiming to restitute the functioning of reality "as such," whereas with contemporary results it goes so far as to presenting itself as technically dependent upon the models which account for the phenomena under study. Causes, in this sense, become interactions and these interactions themselves constitute the fabric of the universe of their manifestations, its geometry: modifying this fabric appears to cause the interactions to change; changing the interactions modifies the fabric.

These considerations may extend to theories of the critical type in that, among other things, the spontaneous breaking of symmetry partially subvert Curie's principle which stated that the symmetry of causes were to be found in those of the effects. And if we want to generalize the Curie principle to the case of the effects which manifest symmetry breaking relative to causes, then we are led to consider not only a singular experience which manifests this breaking, but the entire class of equivalent experiences and their results. As breakings of symmetry are singularized at random by fluctuations which orient them (this is the fundamental hypothesis which enables us to do without the existence of other "causes" which would not have been considered in the problem), the accounting for all possible experiences contributes in eliminating this random character by averaging it and contributes to restore the symmetry of potentialities in actuality and in average. It nevertheless remains that the precise predictability of the result of a given experiment remains limited: if we know that it necessarily belongs to the class of symmetries authorized by the Curie principle, we do not know which possibility it actualizes, that precisely in which the symmetry finds itself to be spontaneously broken and the usual causality jeopardized.[10]

To conclude, it therefore appears that, at the same time as causal laws are replaced in their theoretical fecundity and their explanatory scope by analyses of "geometric" structures and transformations, it is the very concept of causality which regains its status of regulatory concept and which moves away from the constitutive role we have more or less consciously conferred to it. To return to the remarks we have already formulated according to a different perspective, if the phenomenal explanation and the description of *sensible manifestations* continue in physics to have recourse to the concept of cause, on the other hand, the *constitution* of identity and of physical objectivity is increasingly related to the mathematical structures which theorize them and now confer upon them predictive power. This provides renewed topicality to the considerations voiced by Weyl and his friends (such as my interlocutor).

1.3.6 *Towards the "cognitive subject"*

Let's return to another aspect of Longo's "Mathematical concepts and physical objects." It contains remarks which may constitute the basis of a novel inquiry into human cognition. Let's report two of these indications which deserve further reflections. He writes: *"Foundational analysis, in mathematics and in physics, must propose a scientific analysis of the cognitive*

---

[10]Note that the spontaneous symmetry breaking, in classical frames, has a random origin, which, in contrast to what happens with intrinsic quantum fluctuations, does not throw causality as such into question, but concerns only observability (it is epistemic).

subject and then highlight the objectivity of the construction of knowledge within its reference frameworks or systems". And further on, from a more specific angle: *"The project of a cognitive analysis of the foundations of mathematics requires an explanation of the cognitive subject, as living unity of body and mind, living within intersubjectivity and within history. This subject who traces on the phenomenal veil objects and structures, spaces and concepts which are common to mathematics and to physics, lays outlines which are not 'already there,' but which result from the interaction between ourselves and the world."*

This approach, in a way, takes the direction opposite to Boole, Frege, and their logicist and formalist upholders who wanted to find in logic and formalism "the laws of thought," well, at least those governing mathematical thinking. Without denying the advances enabled by the developments within these fields, Longo proposes on the other hand to see and find essential elements enabling the characterization and objective analysis of the cognitive subject in the development of the practice and in mathematical structures themselves. He does not base this proposition solely upon mathematical rigor and demonstrative capacities (prevalence of proof principles), but also on the conceptual stability they authorize as well as on their aptitude to objectively thematize and categorize the quasi-kinesthetic relationships to experiences as primitive as those related to movement, to space, to order (role of construction principles). Beyond their spontaneous language-related apprehensions (which we know to vary according to culture and civilization), inquiries on the basis of the analysis of "behaviors" and mathematical approaches would enable us in a way to stabilize these experiences and intuitions by abstracting them, by objectivizing and universalizing them, and also by revealing the deepest of cognitive workings which animate them.

Chapter 2

Incompleteness and Indetermination in Mathematics and Physics

Introduction

To continue the preceding discussion, this chapter addresses more specific mathematical and physical topics, which, although being relatively technical, present important philosophical issues. In particular, we will examine the play between construction principles and proof principles. In fact, the epistemological issue related to the completeness of formal theories and of physical theories, as well as the "uncertainty principle" in quantum mechanics, is extremely relevant for the scientific and foundational debate and it relates to that play.

Thus, we will more closely try to understand to what extent the fact of conceiving epistemology as the analysis of a genesis presents the traits of a foundational analysis, in mathematics and in physics. Since Frege, as we have seen, many have attempted to disjoin the analysis of foundations, essentially aiming to wrap mathematics in a fabric of absolute logical certitudes, from the analysis of "origins" in the sense that can be found in (Husserl, 1933), for instance. On the contrary, we aim to reaffirm the "foundational" character of analysis within a human process, that is within those actions in the world, which ground and constitute knowledge. As we said, it is necessary to overcome the traditional distinction between *foundations* and *reconstruction of a genesis*, in Husserl's sense, because scientific certainty and objectivity, hence its foundation, resides first in the capacity to explain the choices and methods of knowledge to which one has recourse, including their historical-constitutive dynamic. Or, the physicist would say, to make explicit the reference system (physical and conceptual), the instruments of measurement and of access to reality. To achieve the construction of objectivities in science, it is necessary to know how to set and to explain these parameters on both the conceptual level as well as on the level of praxis, both empirical and deductive. More than any other discipline, quantum physics, to which we will return in part 2.2, shows us the importance of this explanation as well as the paradigmatic role it may play for the foundation of scientific knowledge.

2.1 The Cognitive Foundations of Mathematics: Human Gestures in Proofs and Mathematical Incompleteness of Formalisms

2.1.1 *Introduction*

The foundational analysis of mathematics has been strictly linked to, and often originated, philosophies of knowledge. Since Plato and Aristotle, to Saint Augustin and Descartes, Leibniz, Kant, Husserl, and Wittgenstein, analyses of human knowledge have been largely indebted to insights into mathematics, its proof methods and its conceptual constructions. In our opinion this is due to the grounding of mathematics in basic forms of knowledge, in particular as constitutive elements of our active relation to space and time; for others, the same logic underlies mathematics as well as general reasoning, while emerging more clearly and soundly in mathematical practices. In this text we will focus on some "geometric" judgments, which ground proofs and concepts of mathematics in cognitive experiences. They are "images," in the broad sense of mental constructions of a figurative nature: we will largely refer to the well-ordering of integer numbers (they appear to our constructed imagination as spaced and ordered, one after the other, along a potentially infinite line) and to the shared image of the widthless continuous line, an abstracted trajectory, as an invariant of the practice of action in space and in time.

Their cognitive origin, possibly pre-human, will be hinted at, while focusing on their complexity and elementarity, as well as on their foundational role in mathematics. The analysis will often refer to Châtelet (1993): his approach and his notion of "mathematical gesture" (in short: a mental/bodily image of/for action, see below) has partly inspired this chapter. However, on one hand, we will try to address the cognitive origins of conceptual gestures, this being well beyond Châtelet's project; on the other, we will apply this concept to an analysis of recent "concrete" mathematical incompleteness results for logical formalisms. A critique of formalism and logicism in the foundations of mathematics is an essential component of our epistemological approach; this approach is based on a genealogical analysis that stresses the role of cognition and history in the foundations of mathematics, which we consider to be a co-constituted tool in our effort for making the world intelligible.

2.1.2 Machines, body, and rationality

For one hundred years, hordes of finite sequences of signs with no signification have haunted the spaces of the foundations of mathematics and cognition and indeed the spaces of rationality. Rules, which are finite sequences of finite sequences of signs as well, transform these sequences into other sequences with no signification. Perfect and certain, they are supposed to transform the rational into the rational and stand as a paradigm of rationality, since *human rationality* is in machines. "Sequence-matching" reigns undisputed: when a sequence of meaningless signs matches perfectly with the sequence in the premise of one of the rules (the first at hand, like in Turing machines), it is transformed into its logic-formal consequences, the sequence in the next line; this is the mechanical-elementary step of computation and of reasoning. This step is certain since it is "out of us"; its certainty does not depend on our action in the world, it is due to its potential or effective mechanizability.

All of that is quite great when one thinks of the transmission and elaboration of digital data, but, as for the foundations of mathematics (and knowledge), a schizophrenic attitude is repeating itself. Man, who has invented the wheel, excited by his genius invention, has probably once claimed "my movement, the movement, is there, *in the wheel* ... the wheel is complete: I can go anywhere with it" (well, the wheel is great but as soon as there is a stair ...); in this way, he thought he could place, or rediscover, his own movement out of himself. In this way, the lever and the catapult have become the paradigm of the arm and its action (Aristotle). Gears and clocks' strings coincide with body mechanisms, including brain mechanisms (Descartes); and the contraction of muscles is like the contraction of wet strings (Cartesian iatro-mechanics in the XVIIth century, see Canguilhem, 2000).

But some claim the last machine, the computer, has been invented by referring to human beings and their thinking, while it was not the case for the lever and the catapult or the clock mechanisms and their springs. These are not similar to our own body parts, they have not been designed as a model of them. Here is the strong argument of formalists: this time the mathematical proof (Peano, Hilbert), in fact rationality, has been first transferred into something potentially mechanic. Then, on this basis, engineers have produced machines. This is a strong argument, from a historical point of view, but it leaves this functionality of human being (rationality, mathematical proof), his intelligence, out of himself, out of his body, his brain, his real-life experience. And schizophrenia remains. It has only preceded (and allowed) the invention of the machine, of computer: it is upstream, in the formalist paradigm of mathematical deduction, indeed cognition, since man, in the minimal (elementary and simple) gesture of thinking, would be supposed to transform finite sequences with no signification into finite sequences with no signification, by sequence matching and replacement (from Peano to Turing, see Longo (2007); some refer to Hobbes and Leibniz as well).

But where is this "human computer" whose elementary action of thinking would be so simple? As for the brain, the activity of the least neuron is immensely complex: neurons have very different sizes, they trigger large biochemical cascades inside and outside their cellular structure, their shapes and the form of their electrostatic field change; moreover, their activity is never isolated from a network, from a context of signification, from the world. In fact, as for thinking, when we go from one sentence to another one, with the simplest deduction ("if ... then ... "), we are not doing any sequence matching, but we move and deform huge networks of signification. Machine, with its very simple logical gates, with its software built on even more simple primitives, can only try to functionally imitate our cognitive activities. Machines do not "model" these activities, in a physical-mathematical understanding of modeling – which means to propose a mathematical (and/or artificial) framework able to reproduce what we consider the constitutive principles of the object modeled. But even the functional imitation is easy to recognize, as we will more closely argue in Chapter 5.

On the other hand, some algebraic calculi require purely mechanical processes and are an integral part of mathematics. These calculi, we shall argue, are the death of mathematics, of its meaning and its expressive richness, if they are isolated from contexts of signification and considered as a general paradigm for it.

2.1.3 *Ameba, motivity, and signification*

What else have in common clocks, formalisms, and computers? All artifacts are actually constituted by *elementary* and *simple* components: washers and ropes, 0 and 1 sequences, logical gates, all individually very simple, are put together and associated into huge constructions that may reach a very high complexity. In fact, complexity is the result of a construction which superposes extremely simple constitutive elements: this ability to be reproduced, to be accessible (that is, to be dismantled component by component) is the strength of artificial constructions, which, by this, always proceed "bottom-up." On the contrary, the *elementary* biological component, the cell, is extremely complex; it contains all the objective complexity of life; it is elementary since once the cell is cut, it is not living anymore. And embryogenesis is always "top-down."

The contraposition between artificial and natural, just sketched above by referring to the elementary and simple aspects of artifact as opposed to the elementary and complex components of natural phenomena is the core of our analysis: we will face it again in the complexity of linguistic symbols, of living cells, of strings in quantum physics. At each phenomenal level of this three level classification (human languages, biological entities, microphysics), rough yet historically rich, of the way the world appears to us, the elementary seems extremely complex, perhaps the most complex in the phenomenality and for sure the most difficult to understand.

Moreover, a cell like the ameba or the paramecium changes internally as well as in its relationships with the external: it moves. This is essential for life, from its action in space to cognitive phenomena since "motivity is the original intentionality" (Merleau-Ponty, 1945). Now, in our opinion, signification is constituted by the interference of signal with an intentional gesture, be this gesture "original" or not. In this way, gesture, which begins in motor action, set the roots of signification between the world and us, at the interface of both. The chemical, thermal signal, which affects ameba and cell, is "significant" for the living, regarding its current internal change, its action, and its movement. The neuron, reached by a synaptic discharge which deforms its membrane and its electrostatic field, reacts with a biochemical cascade, with a subsequent deformation of its electrostatic field, even by changing form and place of synaptic connections. In other words: it reacts with an action, a gesture at its scale, with its internal and external mobility; at its level, this reaction is meaning. And the elementary, minimal, living unit is preserved while the current action is modified by the signal (each neuron, as any living thing, is always acting). This modification is at the roots of signification. Of course, the neural network also changes and the net of networks, and the brain as well, in a changing body. This is the modified activity of this entanglement and the awfully complex coupling of organization levels which makes significant the friction between the living and the external world. The result is non-additive, indeed non-compositional (it is not possible to rebuild it by assembling elementary significations "piece by piece"), since it requires the activity of the whole network. Finally, perception itself is equivalent to the difference between an active forecast and a signal (this analysis has gone from Merleau-Ponty, 1945 up to Berthoz, 1997). For this reason, perception leads to signification: it depends on prevision, which is an action, and accompanies any other action. Perception is the result of interference between a signal and an action or an anticipation. In short, there is no meaning without an ongoing action.

As for man, signification includes also action within a communicating community, interaction with others in a symbolic culture, rich of language, gestures, and evocations. Thus, intentionality allows signification, on the basis of original intentionality: motion. This very intentionality which envelops any object of thought, as an "aim."

2.1.4 *The abstract and the symbolic; the rigor*

Through this constitutive route, entangled and complex in its origin, which lies on the action of the living, humanity has arrived to propose our symbolic culture. Its symbols are significant, they refer to the world, and they are in resonance with the world. Each symbol, each sentence has a huge "correlation length" in the space of present and history. The correlation space is almost a physical fact: that is why we use this term borrowed from physics. This length describes the possible distance of causal links: each word is almost physically correlated, by an individual or a whole community, to a huge set of words, acts, gestures, and real-life experiences. These links constitute a "manifold" in space and time, in a mathematical sense of manifold, since it is possible to give a geometric structure to it (as has been already undertaken, in the modern analyses of signification spaces, see Victorri, 2002). There is no metaphysics of the ineffable, but rather a concrete, material and symbolic reality of phylogenetic, ontogenetic, and cultural complexity of the human being and his language (see also Cadiot and Visetti, 2001). This takes us far away from "thought as a formal computation." Yet, mathematics is *symbolic*, *abstract*, and *rigorous*.

A symbol is thus a synthetic expression of signification links. It can be realized in a linguistic sign or a gesture, as it is a movement and a body posture. Both may be elementary, as they are the minimal components of human expression, but they are very complex since each meaningful sign, each gesture, and in fact each body posture, is the result of a very long evolutionary itinerary and synthesizes it. Once more then, the elementary component of the symbolic construction, the minimal, meaningful symbol of intersubjective communication, is very complex. And mathematics is *symbolic* and it is meaningful.

In general, communication is both human and animal, since the gesture and the body posture are a part of animal expressiveness. A human linguistic symbol (a word, a sentence, a text) is also an evoked gesture as in the living interaction there is no sign without signification. But this signification relies on relevance and may be multiple: polysemy is in the core of languages and takes part in the richness and expressivity of communication (Fuchs and Victorri, 1996). Is this compatible with mathematics? We will go back to this point later.

But mathematics is also *abstract*, which is another huge cognitive problem. This abstraction starts from the categorizations of reality, proper to every animal neural system (see Edelman and Tononi (2000)), which are based on the independence regarding sensory modality, indeed on multimodality of the sensory-motor loop (Berthoz, 1997). Concepts of our cultures are the organized expression of these categorizations. This expression has been built through an intersubjective exchange within language, which stabilizes experience and common categorization. Mathematical abstraction is a part of this, but it has its own character due to the maximality of its practical and historical invariance and stability.

Finally, mathematics is *rigorous*. The rigor of proof has been reached through a difficult practical experience. It probably started in the Greek agora, where coherence of reasoning used to lead the political debate, where a sketching of democracy has given rise to science, in particular to mathematics, as the maximal place of convincing reasoning. Rigor lies in the stability of proof, in its regularities that can be iterated. Mathematical logic is a set of *proof invariants*, a set of structures that are "preserved" from one proof to another or which are preserved by proof transformations. Logic does not precede mathematics, rather it has followed mathematics: it is the result of a distilled praxis, the praxis of proof. Logic is in the structure of mathematical arguments, it is made of their maximally stable regularities. Of course it has been necessary to distil it from a practical experience not always perfect: the richness and confusion of a large part of mathematics in the XIXth century is an example of this imperfection. Norms were necessary: people did not know what giving a good definition could look like: some were defining their concepts soundly (Weierstrass), but others, not less great, used to confuse uniform continuity with continuity, say (Cauchy). The formalist answer, identifying rigor with *formal* rigor, might have been necessary. Only nowadays are we able to highlight logic in the geometric structure of proofs (Girard, 2001) and to keep away from sequence-matching, a mechanical superposition of sequences of signs which is the motor of any formalism.[1]

Thus, mathematics is *symbolic*, *abstract*, and *rigorous*, as are many forms of knowledge and human exchange. But it is something unique in human communication, based on these three properties, since it is the *place of maximal conceptual stability and invariance*. This means that no other form of human expression is more stable and invariant regarding the transformation of meaning and discourse. In mathematics, once a definition is given, it remains. In a given context, stability forbids polysemy but not meaning. Invariance is imposed on proofs. This can even constitute a definition of mathematics: as soon as an expression is maximally stable and conceptually invariant, it is mathematics. But let us be careful, we use 'maximal' rather than 'maximum' because we aim to avoid any absolute. Moreover, mathematics is a part of human communication and of the tools man has found in order to organize his environment and make it more intelligible.

In order to escape from the rich confusion of the XIXth century, from the "wildest visions of delirium" proposed by the models of non-Euclidean theories (Frege, 1884, p. 20), and from some minor linguistic antinomies, a strong, perhaps too strong, paradigm was necessary in order to establish robust foundations. For the Hilbertian school it was the paradigm of finitary arithmetic as the essence of a logical or formal system. It was a brave

---

[1] For example, a formalist interpretation and a computer use of "*modus ponens*," *from* A *and* (A *implies* B) *deduce* B, that is its "operational semantics," consists only in controlling in a mechanic way that the finite sequence of signs (0 and 1), which codes for the first A, is identical to the sequence which codes for the second A. Following this, "*then*" the machine writes, actually copies, B: this "sequence-matching" (unsuitably called "pattern-matching" in computer science: there is no general pattern, here, just sequences) is all what a digital computer can do, modulo very simple "syntactic unification" procedures. Within recent systems by Girard, see references, only the geometric structure of deduction is preserved (in this case a "plug-in" or a "connection"). This geometric proof structure does not separate syntax from semantics and allows management, like human reasoning, of signification networks.

response to the disorder and to the conceptual and practical richness of mathematics of that time. But a reification, almost a parody, of what is maximally symbolic, abstract, and rigorous: mathematics. A caricature of these three pillars of human cognition: the symbolic, the abstract, and the rigorous. These three notions, very different from each other, have been identified by formalism with each other and with another notion, quite flat, the *formal* as a "finite sequence of signs without signification, manipulated by possibly mechanizable rules." Such sequences are thus manipulated by very simple formal rules, as the rule that checks and changes a 0 for a 1 (or the opposite), one by one, as in a Turing machine (sequence matching and replacement). Thus the identity

$$symbolic = abstract = rigorous = formal$$

constitutes, from the point of view of human cognition, the crowning glory of simplistic thought, indeed just a thought of mechanics.

2.1.5 *From the Platonist response to action and gesture*

Of course, some did not put up with this reifying schizophrenic attitude, the formalism and its fantastic machines, which would split our thought from the locus of certitude, the logico-formal machine. First of all, Gödel, but almost all the major mathematicians who have come after him, from MacLane and Wigner to René Thom or Alain Connes (von Neumann might be the only exception) have reacted by adopting a more or less naive Platonism. The concepts and structure of mathematics are "already there," we just have to discover them, to see them (in fact, such great scholars can "see" these concepts and structures, before doing any proof; this belongs to the practical experience of every mathematician and we should study this cognitive performance as such, without transforming it into any ontology). Sometimes, arguments have referred to the first Gödel incompleteness theorem. This was accompanied with a severe misunderstanding of the theorem and has triggered a kind of "gödelite" which has not disappeared yet from the pathologies of the philosophy of mathematics (Longo, 1999b, 2002).

We should now rediscover the meaning in proofs, in the act of deduction, as it is the aim of Platonists too, but in our case we aim to do so without any pre-existing ontology.

We should rebuild the constitutive background of mathematics, by carrying out an analysis of its cognitive foundations on the basis of the actions and gestures which give rise to signification in humans, as we have just sketched out. This approach will obviously replace, in mathematics and in natural science, the notion of "ontological truth" by *knowledge construction*, the ultimate result of the human cognitive activity, as well as, thanks to this activity on reality, the notion of *construction of objectivity*. Our undertaking is a part of such a project; in the case of proof, Girard's approach is at the forefront of this project. Our background in lambda-calculus and constructive proofs (Girard *et al.*, 1990) has played an important role in the approach we have chosen. Despite its formal origins, lambda-calculus codes proofs in a very structured way: it preserves the organization of proofs and as such differs very much from the coding necessary to reduce formal deduction to the elementary steps computed by Turing machines, as well as by other systems for computability. This coding destroys the architecture of deduction and makes it lose its signification.[2] Girard has proposed his ideas, which highlight the geometry of proof and the regularities that manage and transform meaning, on the basis of lambda-calculus (see Girard, 2001). However, the shade of the formalist methods and philosophies of Church and Curry (Seldin, 1980) is now very far away. Such a geometry of proof is thus rigorous and it is compatible with a cognitive analysis of the constitution of concepts and of the abstract and symbolic structures of mathematics we have proposed. Now, first of all, these structures organize space and time.

As for the gesture, as it is an elementary, yet complex, action of living, it is at the origin of our relation to space, our attempts to organize it, therefore at the origin of geometry. In this view, it is worth reading Poincaré: gesture and movement "evaluate distance" (see the many quotations from Poincaré in Berthoz (1997)). We will focus now on a very ancient gesture, the conscious eye saccade (jerk), which draws the predator chase line (Berthoz, 1997), and on how this may contribute to establishing a mathematical invariant.

2.1.5.1 *The mathematical continuous line*

The example discussed here refers to a constitutive gestalt of mathematics: the line without thickness (without width, as Euclid would put it). From

---

[2]Writing rules of typed lambda-terms are exactly rules of formal deduction. Although everything remains formal, a term allows us to see proof structure showing through (Girard *et al.*, 1990). Böhm trees (even Levy-Longo trees, a slight variation of them, see Longo (1983)) bring a structure to lambda-terms which would be otherwise too flat. Links between types and objects of "geometric categories" (in particular, topos) complete the mathematical richness of system (see Asperti and Longo (1991).

the cognitive point of view, one can refer, first of all and simultaneously, to the role of:

- the jerk (saccade) which precedes the prey,
- the vestibular line (the one that helps to memorize and to continue the inertial movement),
- the visual line (which includes the direction detected and anticipated by the primary cortex).

The isomorphism proposed by Bernard Teissier (of Poincaré–Berthoz, according to his definition, (Teissier, 2005)) is one between the latter two cognitive experiences: action and movement impose (they make us perform) an identification (an isomorphism) between the experience of inertial movement, conducted in a straight line controlled by the vestibular system, and the forward saccade, which precedes the movement. This isomorphism is to be extended by ocular pursuit, as saccades, also being an action, and it has for result, as a pre-conceptual practice, a pure direction, without thickness. The invariant of these three cognitive practices is an outlined line, a pre-conceptual abstraction, an abstract practice: it is that which counts, what is in common, distilled in the memory of the action, for the purpose of a new action.

This practice confers meaning and is at the origin of (it enables) the conceptual, linguistic, and historical baggage by which one manages to propose the continuous line, parameterized on the real numbers in Cantor-Dedekind style. To put it in other words, this line without thickness is the pre-conceptual invariant of the mathematical concept, invariant in relation to several active experiences; it is irreducible to only one of them. This invariant is not the concept itself, but it is foundational and is the locus of meaning: we do not understand what is a line, do not manage to conceive of it, to propose it, even in its formal explication, without the perceived gesture, without it being felt, appreciated by the body, through the gesture which was evoked by the first teacher, through his/her drawing on a blackboard, a gesture which leaves a trace, like in our memories.

It is necessary here to emphasize the key role of memory, in one of its most important characteristics: the capacity to forget. The intentional lapse of memory, as a result of an aim (even preconscious, if we accept to broaden Husserlian intentionality), is constitutive of invariance: from the selective role of vision (intentional), an active glance, a palpation by sight (Merleau-Ponty), to reconstruction by memory, which intentionally selects (even unconsciously) that which is important, in view of the action. Until the conceptual construction, it is the capacity to forget that which is not important, in relation to the goals in question, which precedes the explication of the invariant, of that which is stable in comparison to a plurality of actions-perceptions. Memory selects, by forgetting, and, by this, practices and yields invariance.

But how may one prove this? There is so little work even and simply of a "gestaltist" nature in the foundations of mathematics or even in mathematical cognition! Yet, mathematics organizes the world in the manner of a science of structures: points, isolated and non-structured, are derived, for example, as the intersection of two lines, Wittgenstein tells us, while Euclid actually constructs this. In fact, a line is not a set of points. It is a gestalt. One can rebuild it using points (Cantor-Dedekind), but also can without points (in certain topos by Lawvere (Bell, 1998)). And its cognitive foundation and understanding should largely rely on a gestaltist approach.

In conclusion, the memory of this gesture is a prior experience toward a very important mathematical abstraction. It is an "abstract" animal experience since it is the *memory of a forecast*, a forecast of a line which is not there, the expected trajectory of a prey. Memory takes it out of its context, by forgetting everything "not important," which is not the object of intentionality, of a conscious or unconscious aim. Memory of a continuous line, since space of movement is connected, a thickness-less line, since the line itself is missing, pure trajectory, this is the pre-conceptual experience of Euclidean lines, as well as of our modern lines which are parameterized on real numbers. Here is one of the constitutive pillars of the knowledge construction we are talking about: it starts from the abstract, the categorizing memory of the predator, its memory of actions in space (indeed of their forecast), and goes up to our abstract, mathematical, and rigorous concept of a continuous line, a parametrized trajectory, which is given within language. But the meaning of this conceptual construction, which organizes space and knowledge, stems from the very first gesture of the predator, from its original intentionality, as it is an action; it stems from its meaningful interaction with the environment. Thus, an established mathematical construction, a trajectory parameterized on real numbers, following Cantor and Dedekind, is meaningful to us since we have this (common) gesture in our constitutive background.

2.1.6 *Intuition, gestures, and the numeric line*

Mathematical intuition precedes and accompanies theories, since it constitutes the profound unity – but this time graspable in action – of a theory; yet, it also follows them, since "understanding [a mathematical theory] is to catch its gesture and to be able to continue it" ((Cavaillès, 1981) quoted in (Châtelet, 1993, p. 31, transl. 2000, p. 9)). Now, intuition may be grounded in gestures, which may evocate images. Indeed, Châtelet as well takes up this role of gesture again: "this concept of gesture seems to us crucial in our approach to the amplifying abstraction of mathematics ... gesture gains amplitude by determining itself ... it envelops before grasping ... [it is a] thought experiment" (Châtelet, 1993, p. 31-32, transl. 2000, p. 9-10). In mathematics, there is a "... talking in the hands ... reserved for the initiates. A philosophy of the physico-mathematical cannot ignore this symbolic practice, which is prior to formalism ..." (Châtelet, 1993, p. 34, transl. 2000, p. 11).

Gesture of imagination should be included in this physical and mathematical intuition as "sense of construction." By making this kind of gesture, a human performs a conceptual experiment: "Archimedes, in his bathtub, imagines that his body is nothing but a gourd of water ... Einstein takes himself for a photon and positions himself on the horizon of velocities" (Châtelet, 1993, p. 36, transl. 2000, p. 12). "Gauss and Riemann ... [conceive] ... a theory on the way of inhabiting the surfaces" (Châtelet, 1993, p. 26). The Gauss and Riemann intrinsic geometry of a curved surface and space is the "delirium" against which the logicist response will fly into a rage (see Tappenden, 1995). Later on, the formalists proposed a solution to this "delirium": it consists of meaningless axiomatic systems, controlled by formal rules, whose coherence relies on the coherence of formal arithmetic into which they can be coded (Hilbert, 1899). Only arithmetic and arithmetical induction (as logical laws or purely formal rules, it depends on the authors) are supposed to be the foundations of mathematics, and are considered to be the unique place for objectivity and certainty. Then monomania has started: indeed, on their own, number theory and induction are very important and are essential components of the foundations of mathematics, the problem comes when considering them as the unique basis for mathematics. This focus on language and "arithmetical laws of thought" has given rise to wonderful digital machines but also to a philosophical and cognitive catastrophe which still remains.

But what is this logico-formal induction, the ultimate law of thought, considered by Frege as logical and meaningful and considered by Peano and Hilbert to be purely formal, indeed as a calculus to be computed by a machine?

In formal terms, once a well formed and expressive enough language is given (we need 0, a successor operator, rules for quantification – for all, written ∀ – and a few others, all simple and elementary), we can write the following rule of induction for a predicate A of this language:

$$\frac{A[0] \ \forall y(A[y] \to A[y+1])}{\forall x A[x]}$$

"Categorical" rule for Peano and Frege (although this term was not in use at that time): this means that number theory is contained in this rule and there is nothing else to say; in other words this rule has only one model. But a true "delirium" is going to show up soon: Lowenheim and Skolem will prove that Peano arithmetic, which should have tallied with integers, has models in all cardinalities! Worst: non-standard models give nightmares to logicists, as they are non-elementary equivalent, in technical terms, a consequence of incompleteness. That is, this Fregean theory of absolutes, the integer numbers, may be interpreted by wild structures of any cardinality, which, moreover, may realize very different properties (but Gödel's theorem is required to prove this). True pathologies of incompleteness, with a weird structure of order (non well-founded, in the non-standard part), they are of no use, except to give alternative proofs of modern incompleteness results (an amusing exercise for the author of these lines, in his young age). On the contrary, if one thinks that the three classes of Riemannian manifolds which modelize the fifth Euclidean axiom and its two possible negations, have all acquired an important physical meaning,[3] then one should revise the Fregean orthodoxy: the delirium is the one of arithmetic axioms, of logical induction and its models but surely not the one of geometry.

But going back to the signification of induction may avoid such delirium. Here is a first idea: "conceive the indefinite (unlimited?) repetition of an act, as soon as this act is possible once" (Poincaré, 1902; p. 41). This act, this iterated gesture performed in space, is the well-order of the potentially infinite sequence of integers. But, what is this sequence? It is the result of an extremely complex constitutive itinerary. It started from the counting of small quantities and the establishing of correlations between small groups of objects or the ranking of certain objects as well, as we share with

---

[3] Classical physics, relativity theory and Hadamard's theory of dynamic flows of a negative curve, respectively.

many animals (Dehaene, 1997). Then, it developed through our ranking, counting, and spatial organizing experiences, which are as old as humans, until there was a huge variety of linguistic expressions. Such expressions have often been devoid of any generality and unable to suggest a general concept, since they were based on individual objects (see Dehaene (1997) and Butterworth (1999) about some kinds of enumeration in peoples with no written languages). Perhaps it is only possible to isolate the concept of number once writing has stabilized thought, although this is not immediate. Sumerians used different notations to refer to 5 or 6 cows and 5 or 6 trees. Should we claim they used to have a general concept of integers? It is doubtful. Sumerians and Egyptians as well, reached much later a uniform notation, independent of the numbered object. Greek mathematics followed, with a beginning of number theory (that is, a quite general, invariant, and stable concept, like that of prime number for instance, which does not depend on representation, i.e. it is invariant with respect to the notation). Yet, a full flavor of our theory of numbers needed the establishment of a truly abstract and uniform notation for numbers, of any size, as in Chinese, Indian, and Arabic cultures.

A further understanding of the mathematical practice of numbers requires us to go back to our discussion on the continuous line. As a matter of fact, starting from a little counting by animals, the SNARC effect described in Dehaene (1997) seems to distribute integer numbers on a mental line. This is a cognitive bulk, which we see, for example, behind an elementary judgment (irreducible to a finitary formalism), like the well-order of the integers (mathematically: "a non-empty generic subset of integers contains a smallest element"), a result of the ordering of numerical practices on a line. Because the structured order of the integers also participates in the direction of the movement, of the gesture which arranges them on a line, this gestalt, which remains in the background, but which contributes to organizing them, to arranging them, to "well-ordering them" towards the infinite, in a highly mathematized conceptual space. Moreover, it is not excluded that in order to grasp the statement of the well-order, so complex albeit elementary, a statement which, in a certain sense, is finitary, one may need control over the whole line, and therefore over projective geometry (or over perspective in painting). Here is this network, constitutive of mathematics, as a structured discipline, which participates in the proof, by necessitating the use of complex gestalts even in elementary number theory. So there we have the immense cognitive (and historical) complexity of an elementary judgment, the well-order of the integers (of which the *ad hoc* reconstructions by large ordinals are also very complex, see next section), which transforms the proof not into a chain of formulas, but into a geometry of meaningful and complex correlations, of references, threads relating the mathematical reasoning to a plurality of acts of experience, conceptual and pre-conceptual, as well as to other already constituted mathematical structures. The strength of reasoning, even its certainty, thus lies in its remarkable stability and in its invariance (not an absolute), as a uniformity of deductive methods which is dynamic (as changing throughout history), but also in the richness of the lattice of connections which hooks it to a whole universe of practices and of knowledge, even pre-mathematical, even non-mathematical.

By this, the concept of integer, reached through language, goes back to space again, since number is an "instruction for action," counting and layout in space: a gesture that organizes mental space, the one of "numeric line" what we all share (Dehaene, 1997). This line may be funny in the layman's imagination: it swings, it is finite ... though this is not the case for the mathematician nor anybody with a (even short) background in mathematics. A mathematician can "see" a discrete, growing numeric line, continuing from left to right (in our culture at least, (Dehaene, 1997)) with no limits. This line thus results in the last step of a long itinerary, which leads us to reconsider in the space, a mental space, the concept of number as a generalized iteration or invariant constituted by an iterated gesture in the space of action. On the other hand, the origin of this concept lies also in a temporal iteration. An idea from Brouwer, creator of intuitionist mathematics, can be used to support this hypothesis: phenomenal time is defined as a discrete sequence of momenta, as a "partition of a moment of life into two distinct things, one giving the place to the other, yet remaining in the memory" (Brouwer, 1948). In this view, the mathematical intuition of the integers' sequence would rely on the subjective and discrete sequence of time; then computation would be the development of a process through a discrete temporality.

In conclusion, the concept of number and the discrete numeric line, which structures it, are the invariants constituted by the plurality of acts, experienced in space and time. These invariants owe their independence, which characterize intersubjectivity (since this is the experience shared with others that gets the most stable), to language and writing. Now, the inspection of the discrete numeric line, an image built in our mental spaces, is a mathematical praxis of a huge complexity; it summarizes a conceptual history which starts from the counting of small quantities, goes up to modern mathematical practice and founds it.

2.1.7 *Mathematical incompleteness of formalisms*

The incompleteness theorems tell us that the structured meaning of number, its numeric line, is elementary, though very complex. More precisely, it is impossible to capture (and break down) with elementary and simple formal axioms, a statement such as "a generic non-empty set of integers has a smallest element" or some of its consequences. Here is the formal incompleteness of formalisms.

In order to understand this mathematical statement, a "geometric judgment of well-order," you should leave this page and, once more, have a look at the discrete and growing numeric line you have in your mind. Hopefully, you can see the well order property: first, isolate a generic non-empty subset (generic means here that it has no specific properties, except possessing an element, and thus it is not necessary to define the set explicitly). Then, since this subset contains a number, it contains also a smallest number ... In order to see it, look at the existing element, call it p, in your non-empty set; then, there surely is a smaller, possibly equal to one, in the set. Look at the smallest, i.e. at the first one within the finite part of the number line that precedes p: can't you see it? It is there, even if you cannot and you do not have to compute it. This is the judgment expressed millions of times, by mathematicians (not logicians, of course) using induction in a proof. Of course, well-order implies formal induction, but is much stronger. In fact, well-order is the "construction principle" at the core of number theory and formal induction, which is a proof principle, and cannot capture it; this is the mathematical incompleteness of formalisms: some formal statements of arithmetic can be deduced by this judgment, but not by formal induction. Generally speaking, it is possible to describe this mathematical incompleteness as a gap between (structural) *construction principles* and (formal) *proof principles*: in short, (well-)order (and symmetries, we will go back to this), as construction principles, allow us to prove more than induction, a proof-principle (the key one, in the formalist-logicist tradition).

Now, it is impossible to understand this story through the proof of Gödel's incompleteness theorem, which is "only" a fantastic undecidability theorem. That is, Gödel's theorem "only" produces a proposition, which is unprovable (as well as its negation). However, this becomes clear in the proofs of more recent concrete incompleteness theorems (see Longo,

2002; this paper includes a technical discussion[4]). Some persist in "forcing" formal induction in order to prove these theorems. It requires a technically extraordinary difficult *ad hoc* construction, which forces induction all along ordinals, far beyond the countable or predicative (see Longo, 2002, for some references). Nevertheless, the only uniform method, which proves these formally unprovable theorems, remains a concrete reference to the numeric line; moreover, this method is included in the background of any non-logician mathematician. Such a mathematician understands and uses induction in the following way: "the set of integers I am considering is non-empty, therefore it includes a smallest element." That's all and it is cast-iron. Once out of the Fregean and formalist anguish, this means out of the forbidden foundational relation to space and out of the myth according to which certainty relies only on sequence matching and formal induction, then any work based on the ordered structure of numbers, on the geometric judgment lying at the core of mathematics, can go smoothly. Incompleteness shows that this judgment is elementary (it cannot be further reduced), but it is still a (very) complex judgment.[5]

---

[4] Let us go back to the above exercise on the numeric line. The computability of the first element in the non-empty set will depend (on the level) of the definability of the considered subset. In some recent "concrete" examples, we prove that, in the course of a proof, we use a subset of integers whose definition, although rigorous, cannot be given in first-order, formal arithmetic, which is the place for effective computability. For this reason, such a first element (if it exists) is far from being computable. But humans (with a background in mathematics of course) can understand very well the conceptual construction and the rigorous, though non formal, proof without any need of an ontological miracle; in such a way, we can prove theorems which are formally unprovable (see the next footnote and Longo (2002) where normalization and Kruskal-Friedman theorems are discussed). If computers and formalist philosophers cannot do nor understand these proofs, this is their problem.

[5] Any strictly formalist approach also rejects some principles that are much less strong than the latter, for such an approach rejects even imperfectly "stratified" (predicative) formal systems: the elementary must be absolutely simple and must not allow any "complexifying loop" (self-reference). But, impredicativity is ubiquitous in the Kruskal-Friedman theorem, KF (see Harrington *et al.*, 1985, in particular Smorinsky's articles). It is the same for normalization theorems in formal, though impredicative, type theory (the F system (Girard *et al.*, 1990), which has played a very important role in computer science). In fact, its proof through formal induction would require a transfinite ordinal, far beyond the conceivable (but the analysts of ordinals are prepared to do anything, except using the geometric judgment of well-order; others, the predicativists, prefer to throw the system itself through the window: such a monstrous ordinal would confirm it is not founded). Another formal analysis of normalization, relevant for the computer-aided proof, prefers to use third-order arithmetic; but ... by which theory is the coherence of the latter guaranteed? By the fourth-order arithmetic and so on (likewise if a formal set theoretic framework is chosen)? In short, the "classic" proof of KF mentioned above, uses, in a crucial deductive passage, the geometric judgment of well-order, regarding

The natural dimension of mathematics, or in other words its blend of artificial and natural, is built on that: mathematics surely includes some completely and uniformly axiomatizable fragments that can be captured by elementary, *simple*, and mechanizable principles (this is the most boring part of mathematics, which is now being transferred to computers); however, mathematics is also based on complex and elementary judgments, such as the geometric judgment of well-order which completes induction and provides its geometric foundations. But does the use of this judgment make the notion of proof undecidable?[6] This is a problem for the machines but not for humans: in fact, the geometric judgment of well-order has a "finite" nature and is quite effective from the point of view of a numeric line (only a finite initial segment is to be considered, though not necessarily computable: this is the segment which precedes an element of the non-empty subset considered). This line belongs to the human mental spaces of conceptual constructions, which are the result of action in space and common linguistic experience and, at the same time, of the (spatial) reconstruction of phenomenal time. It is objective and efficient, as any mathematics, because of its constitutive background which fixes its roots in the relation between the world and us. The "(very) reasonable effectiveness of mathematics" comes from its blend of formal calculi and meaningful naturality.

2.1.8 *Iterations and closures on the horizon*

But what is this finite, so important to formalisms and machines, since it defines the computable, the decidable? Actually, it is not possible to define it formally. Another astonishing consequence of the incompleteness of logico-formal approaches is that there is no formal predicate that could determine the finite without having to determine the infinite too. In short, it is not possible to isolate the collection of standard integers, without an axiom or a predicate for the infinite; in other words, formal arithmetic cannot talk about (standard) finite numbers. To do so it is necessary to use a version of set theory including an axiom for infinite. (An analogous

---

a supposed non-empty and highly non-calculable subset (Σ_1^1 in technical terms). The formal proof of normalization, and the meaningful one as well, uses, *de facto*, the same judgment as the only guarantee of coherence, indeed of sense, and both reject any infinite regression (see (Longo, 2002) for a compared analysis of the provability of these major results of contemporary logic).

[6]It is possible to characterize formal Hilbertian systems, in a very wide sense, as deductive systems in which the notion of proof is decidable, in the sense of Turing machines.

situation exists in category theory, more precisely in toposes with natural number objects.)

Here is another way to understand why it is not possible to overlook the "numeric line" (or an axiom for infinity): the concept of integer is extremely complex, it is necessary to immerse it in a rich structure, the well-ordered sequence of numbers, an infinite set, in order to grasp it fully. Nevertheless, the difference between both kinds of structure is clear: in contrast with our approach, set theory is ontological or formal. In the ontological case, objectivity and certainty are guaranteed by a pre-existing absolute (possibly God, who is certainly, for many, very reliable), while in the formal case they depend on the formal coherence of the theory. Now, the only method to formally prove coherence of arithmetic, the theory of finite numbers, is a proof done by transfinite induction or within the framework of a formal set theory including an axiom of infinity, for a larger infinite cardinal number ... thus, the formal coherence game is going on" *in perpetuum*," as detached, regarding the world, as Platonist ontology. This procedure, which needs to be iterated to higher infinities to prove the consistency of transfinite induction or of set theories extended by axioms of infinity, provides useful classifications of theories in terms of "proof-theoretic strength," but have little epistemological relevance. In contrast, the geometric judgment of well-order we focused on, is based on a phenomenal real life experience which has started out of mathematics and in particular out of number theory: the foundations of mathematics are provided by the cognitive origin of this judgment, by its phylogenetic history based on the plurality of our modes of access to environment, space and time within the framework of intersubjectivity and language. This judgement is to be added to formal proofs of coherence, which are sometimes very informative, like normalization theorems (see Girard *et al.*, 1990, Longo, 2002) , and allows us to stop infinite foundational regressions.

Therefore, within mathematics it is formally impossible to avoid using the infinite in order to talk about the finite. In reference to terms from physics, we could say that finite and infinite are formally entangled. But the current infinite is a "horizon": we can understand it as a limit to the numeric line, as the vanishing point of projective geometry or of the paintings of Piero della Francesca who is one of its creators. Châtelet formulates this very nicely: "With the horizon, the infinite at lasts finds a coupling place with the finite" (Châtelet, 1993, transl. 2000, p. 50) ... "An iteration deprived of horizon must give up making use of the envelopment of things" (Châtelet, 1993, transl. 2000, p. 52) ... "Any timidity in deciding the horizon tips the infinite into the indefinite ... It is therefore necessary, in order to refuse any concession to the indefinite and to appropriate a geometric infinite, to decide the horizon" (Châtelet, 1993, transl. 2000, p. 52). This is what Piero has done by perspective in his paintings and what mathematicians, at least since Newton and Cantor, do everyday. In mathematics, the infinite is a tool to understand better the finite: since Piero and Newton, we better understand finite space, around us, and movements in it (speed, acceleration ...), by the infinite.

In contrast, a machine iterates, since "finitude fetishizes iteration" (Châtelet, 1993, transl. 2000 p. 51): one operation per nanosecond without any weariness nor boredom. Here is the difference since, in such situation, humans (and animals) are bored. After a couple of iterations, we get tired and we stop or say: "OK, I got it" and we look at the horizon. This is the true "Turing test," as boredom should be added in order to better test the human-machine difference (see Longo, 2007 for other arguments).

Intermezzo: young Gauss and induction

At the age of seven or eight, Gauss was asked to produce the result of the sum of the first n integers (or, perhaps, the question was slightly less general ...). He then proved a theorem, by the following method:

1	2	...	n
n	(n-1)	...	1
(n+1)	(n+1)	...	(n+1)

which gives $\Sigma_1^n i = n(n+1)/2$.

This proof is not by induction. Given n, a uniform argument is proposed, which works for any integer n. Following Herbrand, we will call this kind of proof a **prototype**. Of course, once the formula is known, it is very easy to prove it by induction, as well. But, one must know the formula, or, more generally, the "induction load."

As a matter of fact, little Gauss did not know the formula, he had to construct it as a result of the proof. And here comes the belief induced by the formalist myth: proving a theorem is proving an *already given formula*! This is what we learn, more or less implicitly, from the formal approach: use the axioms to prove a given formula. An incomplete foundation and a parody of mathematical theorem proving. Note that, except for a few easy cases, even when the formula to be proved is already given (the most known example: Fermat's last theorem), the proof requires the invention of an induction load and of a novel deductive path which may be very far from the formula (in Fermat's example, the detour requires the construction of an extraordinary amount of new mathematics). This is well known also in automatic theorem proving, where human intervention is required even in inductive proofs, as, except for a few trivial cases, the assumption required in the inductive step (the induction load) may be much stronger than the thesis, or with no trivial relation to it. Clearly, *a posteriori*, the induction load may be generally described within the formalism, but its "choice," out of infinitely many possible ones, may require some external heuristics (typically: analogies, symmetries, symmetry breaking). More generally, *proving a theorem is answering a question*, like Gauss's teacher's question, concerning a property of a mathematical structure or relating different structures, *it is not proving an already given formula*. Let's speculate now on a possible way to derive Gauss's proof. In this case, little Gauss "saw" the discrete number line, as we all do, well-ordered from left to right. But then he had a typical hit of mathematical genius: he dared to invert it, to force it to go backwards in his mind, an amazing step. This is a paradigmatic mathematical invention: constructing a new symmetry, in this case by an audacious space rotation. And this reflection symmetry gives the equality of the vertical sums. The rest is obvious. In this case, order and symmetries both *produce* and *found* Gauss's proof. Even *a posteriori*, the proof cannot be found on formal induction, as this would assume the knowledge of the formula.

2.1.9 *Intuition*

So far, we have discussed little about intuition, but Gauss's example forces us to go back to its analysis. This word is too rich in history for it to be easily dealt with. Too often, "balayé sous le tapis" it ends in the black holes of explanation. Rigor has been quite fairly opposed to it, up to the *"rigor mortis"* of formal systems. Many errors in proofs, in particular during the XIXth century, have justified such a process (we already mentioned Cauchy's mistakes; but also Poincaré's in his first version of the three bodies theorem should be quoted; others used to "look" at the continuous functions and to claim they were all differentiable, from right or left as Poinsot). But especially, non-Euclidean geometries' delirium had broken "*a priori*" geometric intuition, the ultimate foundation of Newton's absolute Euclidean spaces, in their Cartesian coordinates.

Now, the foundational analysis of mathematics we are developing does not imply the acceptation of whatever "intuitive view." On the contrary, selection must be rigorous, and justification must propose a constitutive analysis of a structure or a concept. In fact, intuition itself is the result of a process which precedes and follows conceptual construction; intuition is dynamic, it is rich in history. After Cantor, for example, a mathematician cannot have the same intuition of the phenomenal continuum as before: he even has some difficulties to view it in a non-Cantorian way.

The dialog with the sciences of life and cognition allows us to reduce the reference to introspection, which was the only tool used by previous analysis in this direction (Poincaré, Enriques). Like any scientific approach, our analysis tells a possible constitutive story, to be confirmed, to be refuted or to be revised: any knowledge, any science, must be strong, motivated, and methodical: however, it remains as uncertain as any human undertaking. Cognitive science analyses, and by doing so, calls into question the very tools of thought: it must then propose a scientific approach, which would be the opposite of a search for certainty lying in the absolute (i.e. non-scientific) laws of logicism.

The examples which have been proposed here, well-ordering and the continuous line as a Gestalt, could be considered as paradigmatic, because of their important differences. The reflection, which has been carried out above, about geometric judgment of well order (discrete numeric line), is based on one century of work on arithmetic induction and follows the quoted cognitive analyses (Dehaene, 1997; Butterworth, 1999; "Mathematical Cognition", which focuses on numeric deficits and performances has become a discipline and a journal). The essential incompleteness of formal induction thus indicates the huge mathematical soundness of a common praxis which is an aspect of proofs. As we have said, this praxis is a part of proofs, since, when induction on ordinals or orders (of variables) is forced, it always requires using a further ordinal or a higher order whose justification is not less doubtful. Descriptions of ordinals and orders within set theory similarly leads to an infinite piling-up of the absolute universes we just talked about. On the other hand, the infinite conceptual regression stops at our geometric judgment of well order: if one wants to know what is going on in number theory, there is, at the moment, no other way. This corresponds to the feeling of any mathematician and this is expressed, usually, by a Platonist attitude: the number line is there, God given. Let us change then this pre-existing ontology, concepts without human conceptor, for an analysis of human construction of knowledge.

Poincaré and Brouwer may have opened the way, but the technical developments have followed only the ideas of the latter. However, these developments have undergone, on the one hand, a complete loss of any sense because of the formalization of intuitionist logic by Heyting and his successors (see Troelstra, 1973) and, on the other hand, by the philosophical impasse of the Brouwerian solipsism and language-less mathematics (see van Dalen, 1991): such ideas are completely opposed to the constitutive analysis we propose here, which refers, in a fundamental way, to the stabilization of concepts occurring within shared praxis and language in a human communication context. After one hundred years of reflections on the topic, mathematical experience and cognitive analyses suggest how to go back to the practice of inductive proofs and how to use it as a cognitively justified foundation of deduction. In this case, intuition comes at the end of a process which includes also a practice of proof; in short, intuition follows the construction of the discrete numeric line and allows (and justifies) geometric judgment.

The other example introduced above, the memory of a continuous trajectory, uses some ideas sketched in Longo (1997 and 1999) and comes from recent remarks in cognitive science (neurophysiology of eye saccades and pursuit in Berthoz (1997)). However, it does not propose any foundation to proof. It consists just of a reference to signification which precedes and justifies conceptual construction, on the basis of its pre-conceptual origin. From Euclid's viewpoint on the phenomenal continuum to Cantor and Dedekind's rigorous construction, mathematics has managed to propose (to create) a continuous line from the abstract trajectory already practiced in the human activities and imagination: in this case, intuition precedes mathematical structure and then is enriched and made more precise by the latter. A mathematician understands and communicates to the student what the continuum is by gesture, since "behind" the gesture both share this ancient act of life experience: the eye saccade, the movement of the hand. With gestures and words, a teacher can (and must) introduce to "the talking in the hands ... reserved for initiates" what Châtelet talks about. The conceptual and rigorous re-construction is obviously necessary: the one of Cantor and Dedekind is one possible example (see the works of Veronese around the late XIXth century or Bell (1998) for different approaches), but teaching must also make the student feel the experience of intuition, the experience of "seeing," which lies at the core of any scientific practice.

In this example, the original intuition may not be essential to proofs, however, it is essential to comprehension and communication and in particular to conjecture and the construction of new structures. Indeed, here lies the serious lack of logicism and formalism, which are entirely focused on deduction: the analysis of the foundations of mathematics is not only a problem of proof theory, but it is also necessary to analyze the constitution of concepts and structures. Set theory has accustomed people to an absolute Newtonian universe where everything is already said, one just has to make it come out with the help of axioms. On the contrary, mathematics is an expanding universe, with no "pre-existing space," to which new categories of objects and transformations are always being added. Relative interpretation functors allow for reconstructing dynamic unity or correlations between concepts and structures, new and ancient.

The two examples which have been studied may seem modest. Nevertheless, they may play a paradigmatic role: the concept of integer number and its order, and the continuous structure of a thickness-less, one-dimensional line, are two pillars of mathematical construction. As already mentioned, the construction of a point without dimension in Euclid is given as the intersection of two one-dimensional lines or points (semeia, actually) such as at the end of a segment; this is also a way to get to and grasp the even more abstract notion of a point. Of course, other examples should be analyzed: the richness and "open" nature of mathematics requires analyses of greater richness. For example, we should investigate the reconstruction of borders and lines, that often do not exist, as in some experiments of Gestalt theory. This is another pre-conscious practice of a preliminary form of abstraction, probably at the very low level of the primary visual cortex. From "Kanizsa triangles" and many other analyses (see Rosenthal and Visetti, 1999) to works in neurogeometry (see Petitot, 2008) we are going to understand the richness of the activity of visual (re-)construction. Since vision is far from being a passive perception, it is rather a "palpation through looking" (Merleau-Ponty), it participates in a structuring and a permanent organizing of the environment. We extract, impose, project ... forms, which is a kind of pre-mathematical activity we share with at least all animals that are equipped with a fovea and a visual cortex (almost) as complex as ours. The friction between us and the world produces mathematical structures, from these elementary (but often complex) activities, up to language and concepts, as soon as this "friction" involves a communicating community. Within it, writing further stabilizes the conceptual invariant.

The analysis of proof also requires this kind of investigation, since no proof of importance comes without the construction of a new concept or a new structure. And this is what is really worthwhile in mathematics. Besides the incompleteness of arithmetic, the one of formal set theory, to which statements such as the continuum hypothesis escape, shows that even *a posteriori* formal reconstruction is often impossible. Finally, the scientific approach to intuition should also be built in order to think about the teaching of mathematics. The usual way to teach mathematics, as "an application of (formal) rules," is a punishment for any student and may have helped the current decrease in mathematical vocations. It is urgent to go back to signification, to the motivated construction, in order to recover and communicate the pleasure of mathematical gesture.

2.1.10 *Body gestures and the "cogito"*

Complex gesture (which may be non-elementary) evocated by Châtelet, helps us to understand what is at stake. However, it is necessary to "naturalize" these gestures much more than what has been done by Châtelet. Indeed, a limit to his thought lies in a refusal of the animal life experience which precedes our intellectual experience, and also in the lack of an understanding of the biological brain as a part of the body. This body allows gesture among humans, not only in a historical but also in an animal dimension. This is precisely the sense of humanity we need to grasp. The absence of any understanding of human, regarding its natural dimension, is the biggest mistake of the great and rich turn of philosophy, the Cartesian "cogito." This, in turn has formed a gap between us and our animal activities in space and time, and has directly led to the myths of machine or to ontologies exterior to the life world. If only modern philosophy had started with "I am scratching my nose and my head and I think, therefore I am," we would have progressed better. But scratching is not the most important and, provocation aside, what is especially to be grasped is the role of prehension and kinestheses in the constitution of our cogitating human mind, starting with the consciousness of our own body (see Petit, 2004). We further construct the self in its relation with others, up to explicit thinking in language. In such an undertaking, there would be a reference to a wide gesture which makes us conscious of our body and, through action, places this body in space. The reflexivity/circularity of the abstract and symbolic thinking of self would find there its elementary, though very complex, explication, in thinking of a deduction which would have, as a consequence, at once the thought itself, the thinking of self, and the consciousness of being alive within an environment. Husserl describes a strong gesture, another original act of consciousness, the one of a man who "feels one of his hands with the other one" (Petit, 2003; Berthoz and Petit, 2003). This biological, material hand is the very first place for gesture; humanity, actually the human brain, wouldn't exist without the hand. This is not a matter of metaphors but rather a concrete reference to what we have been taught by the evolution of species: in the course of evolution, the human free hand has preceded the human brain and has stimulated its development (Gould, 1977). In fact, the nervous system is the result of the complexification of the sensori-motor loop. The human brain is what it is because the human hand is for this loop the richest possible tool, the richest possible animal interaction in the world. Then, socialization and language (because of their complexity and expressivity) and history as well, have done the remaining work, up to mathematics.

What is lacking in formal mechanisms, or in other words their provable incompleteness, is a consequence of this hand gesture which structures space and measures time by using well order. This gesture originates and fixes in action the linguistic construction of mathematics, indeed deduction, and completes its signification.

2.1.11 *Summary and conclusion of part 2.1*

As we said at the very beginning, mathematics has always been an essential component of the theories of knowledge. So there lie, in our opinion, the reasons of this focus, reiterated throughout the course of history: they are due to the anchoring of mathematics in some of the fundamental processes of our interaction with the world. The gesture which traces a trajectory, an edge, the following of a prey by eye jerks (saccades), the memory of these gestures, as well as counting to keep, divide, compare ... are among some of the most ancient acts carried out by the living beings that we are, and they participate in mathematical construction. It even seems that writing began with the quantitative recording of debts, by the Sumerians (see Herrenschmidt, 2007). In short, it may be that the first great conceptual invariants have been proposed in their specifically mathematical maximality, as developments, complex and constituted through human communication, of the most fundamental originary organizing gestures for space and action in space. The cognitive sciences and mathematics have all to gain from a two-way interaction, through the examination of the great problems of the foundations of mathematics.

In particular, we proposed to analyze two features of mathematical reasoning, namely: the construction of mathematical concepts and the structure of mathematical proofs, in order to rediscover sense in mathematics.

The idea is that signification is based on language first, but also on gestures, as forms of action and communication, in a broad sense, it actually originates by interference with action; moreover, we point out that linguistic symbols, which are essential to intersubjectivity as a locus for human abstraction, are grounded as a last resort on gestures. Yet, mathematical concepts and proofs are developed within language, as a need for intersubjectivity in the context of communication; and the linguistic framework brings further stability and invariance to mathematical concepts and proofs, in particular since writing exists.

In our perspective, thus, the signification of concepts and proofs relies also, in contrast with a Platonist and a formalistic view, on some particular features of human cognition. These features precede language, resort from action and ground our meaningful gestures. We gave two examples of them: the complex constitution of the integer number line and of trajectories, from eye saccades to movement. The first grounds the notion of the well-ordered number line, a constructed mental image, on which the principle of induction used in mathematical proofs relies. Similarly, the image of the continuous line/trajectory is the result of a variety of gestures, including eye saccades. It founds and gives meaning to the subsequent mathematical conceptual (linguistic) construction.

The proof itself, as a particular case of deductive reasoning, relies on gestures: this a consequence of the sketchy analysis developed here of recent "concrete" incompleteness theorems, which show exactly where formal induction is insufficient. Thus, gestures may be involved in mathematics and proofs at different levels. In suggesting and grounding the constructions of mathematical structures and in proofs, by completing the principle of induction, as in the example. But also in the deductive structure of the proof itself: the geometry of proof developed by Girard, as a new paradigm of deduction (in contrast with "sequence matching", the tool of formalism) could be used to uncover the organizing structures involved in proofs. In this view, the explicit deductive process, as a result of an exigency of communication, is constrained by language but implicitly involves structured gestures, similarly as concept formation.

The cognitive foundation of fragments of the mathematical practice hinted at here is clearly an attempted epistemological analysis, as a genealogy of praxes and concepts, in language and before language. The historical construction of mental images is a core component of it, as a key link to our relation to space and time and a constitutive part of our ongoing attempt to organize the world, as knowledge.

2.2 Incompleteness, Uncertainty, and Infinity: Differences and Similarities Between Physics and Mathematics

With the view of developing this chapter's theme around the notion of incompleteness, such as it appeared with force during the 1930s, a period which saw the flourishing of Gödel's great logical theorems, and such as it started to foster debates and arouse new perplexities in physics during the same era, we will delve here into the conceptuality of quantum physics (to which results we will also be called to refer to on several occasions). In the following chapter, from a broader point of view, we will attempt to situate this field of physics with regard to the other great theories having marked the XXth century: relativity and the theories of dynamical or critical systems. These theories will prove to be less destabilizing on a conceptual level, but will, nevertheless, contribute in clarifying certain aspects addressed here, in a complementary and relatively simpler fashion.

Firstly, and somewhat trivially, we know that in physics the accumulation of empirical proof does not suffice to account for the totality of the theoretical construction which represents phenomena. This is how, in physics and at a very first level, is manifested the incompleteness of proof principles relative to construction principles (*cf.* Chapter 4 for the constructive role of the geodesic principle). However, this incompleteness has been thought of as way beyond this first level, both in classical and in quantum frames.

2.2.1 *Completeness/incompleteness in physical theories*

2.2.1.1 *From logic to physics*

We first return to a conceptual comparison between the positions of Laplace and Hilbert on the one hand and of Poincaré and Gödel on the other. The first two refer to a sort of strict completeness of theories that are physico-mathematical in one case, and purely mathematical in the other: they would be complete in the sense that any statement concerning the future would be decidable (this is the Laplacian predictability of systems determined by a finite set of equations) or, regarding logico-formal derivations, the completeness or decidability of arithmetic (or of any sufficiently expressive axiomatic, thus finitistic, theory: this is Hilbert's completeness and decidability conjecture). The other two demonstrate incompleteness: Poincaré, with his theorem on the three-body problem, showed unpredictability of interesting non-linear systems, which we may understand as undecidability of future states of the physical system (an explicitly given statement about it would be undecidable or underivable from the equations). Their dynamics will be said to be sensitive to the initial conditions, and in that, deterministic, yet unpredictable: minor variations, possibly below the level of observability, could cause major changes in the evolution of the system. Gödel proved the unprovability of coherence and the intrinsic undecidability and incompleteness of arithmetic (and of all its formal extensions), by constructing an undecidable statement, equivalent to the formal assertion of consistency, which is thus also unprovable within the system.

This analogy here is informal, as Hilbert asked for the decidability of purely mathematical statements, while Poincaré's unpredictability concerns the relation between a system of equations and a physical process. Yet, on one side Poincaré's philosophical critique of Hilbert's program (the "sausage machine of formal mathematics that would automatically prove all theorems") is based on the intuition of the "unsolvable" ("the provably unsolvable problems are the most interesting as they force us to find new ways" he writes in 1908). On the other side, recent results show an asymptotic correspondence between dynamic unpredictability and undecidability. In short, classical physical randomness as deterministic unpredictability may be shown to be equivalent to algorithmic randomness, a strong form of (logical) undecidability: they coincide for infinite sequences or trajectories in effective dynamical spaces (thus the correspondence may be considered asymptotic, see Longo (2009).

2.2.1.2 *From classical to quantum physics*

It can then be interesting to try to characterize the main types of physical theories[7] themselves in terms of this relationship to completeness.

If relativistic theories may indeed proclaim theoretical completeness, in the sense defined above, it very well appears that critical theories on the one hand and quantum theories on the other hand, may present two distinct manners by which they manifest incompleteness.

In the case of chaotic dynamic systems, for example, unpredictability, as we observed, is associated with the sensitivity to the initial conditions joint to the non-linearity, typically, of the equational determination (which we will call formal, see Chapter 5). It may, however, be observed that an

---

[7]It is the distinction between relativistic, critical, and quantum theories which we will detail in Chapter 3.

infinite precision regarding the initial data is meant to generate a perfectly defined evolution (deterministic aspect of the system), or that a reinitialization of the dynamic system with rigorously *identical* values leads to reproducible results. This is *in principle* and from the *mathematical* viewpoint, of course, because physically speaking, we are still within the context of an approximation and the result of a measurement is always, in fact, an interval and not a unique point. Hence, we may notice that an essential *conceptual discontinuity* appears between that which pertains to a finite level of approximation and that which constitutes a singularity, at the "actual" infinite limit of precision (or, one could say, between the unpredictable and the reproducible).

In quantum theories, contrastingly, as we shall see, whether it is an issue of relationships of indetermination or of non-separability, unpredictability is intrinsic to the approach, it is inherent to it. In this case, the degree of approximation matters little: there is a sort of conceptual *continuity* between finite and infinite, the observed and measured behaviors do not present any particular transition. Another way to observe this is by noticing that this time, regardless of the rigor of the reinitialization of the system, the results of the *measurement* will not necessarily be individually reproducible (probabilistic character of quantum measurement), even if the law of probability of these results can be perfectly well known.

It is randomness then which is at the center of these theories, and this by subtle differences from classical frameworks. This point is delicate and we refer the reader to chapter 7 dedicated to this theme.

2.2.1.3 *Incompleteness in quantum physics*

In the debate on the completeness of quantum physics, Einstein, Podolsky, and Rosen (EPR) highlighted three characteristics of a physical object which they believed to be fundamental in order to be able to speak of a *complete* theory (Einstein *et al.* 1935; Bohm, 1951):

- the reference to that which they called elements of *reality* (as something "beyond ourselves");
- the capacity to identify a principle of *causality* (including in the relativist sense);
- the property of *locality* (or of *separability*) of physical objects.

Bell inequalities (Bell, 1964) and their experimental verifications, notably by Aspect and his team (Aspect *et al.*, 1982), have shown that the third EPR postulate was not corroborated: experience shows that two quantum objects having interacted remain for certain measurements a single object, consequently *non separable*, regardless of their distance in space. In short, two quanta having interacted, even if they are afterwards causally separated in space, must be considered as a unique object, any measurement of a value on the one would automatically determine the value of the other.

Presented this way, the eventual *incompleteness* (or, conversely, *completeness*) of quantum physics seems to have nothing to do with what is meant by completeness or incompleteness in logic[8] and in model theory, which have been addressed in length (Löwenheim-Skolem, Gödel, Cohen – the continuum hypothesis, up to the recent results by Friedman-Kruskal and Girard, which we have also addressed). However, a deeper examination reveals that what appears at first glance as a lexical telescoping may not be completely fortuitous. To each of these characteristics "required" by EPR in physics, one may, indeed, without distorting the significations too much, associate characteristics "required" by mathematics:

- elements of reality would be put into correspondence with proofs that construct existence, that is, the effectiveness of mathematical constructions (which, axiomatically or not, and we have seen the difference, cause mathematical structures to exist);
- the principle of causality would be put into correspondence with the effectiveness of the administration of proof (which presents and works upon the rational sequences of demonstrations, be they stemming from formalism as such or not);
- the property of locality (or of separability) would be put into correspondence with the autonomy of mathematical theories and structures inasmuch as they would be "locally" decidable (or that within a formal theory, any statement or its negation would be demonstrable).

Now it is precisely this local autonomy of theories, this "locality" in terms of decidability, which seems to be contradicted by the theorems of incompleteness in mathematics. The latter indeed refer to a sort of globality of mathematical theories in that, for example, the adjunction of axioms to a theory may render it if not "decidable", at least "more expressive" (or

---

[8]By accepting, for this discussion, to use the framework of arithmetic, of set theory (ZFC type), and its models, a framework within which the questions relative to logical completeness were first raised.

capable of deciding previously undecidable statements), but at the cost of generating a new theory which requires the same treatment itself, because, remaining formal, it would still be incomplete.

But we can probably push the analysis further than suggested by these conceptual analogies. If we refer to the interpretation of Gödelian incompleteness in terms of discrepancy between construction principles (structural and significant) and proof principles (formal), see part 2.1, that is, in terms of incomplete covering, of *semantics* by *syntax* (all achievable propositions are not formally derivable, or, more traditionally: semantics exceeds syntax), then a closer relationship may be established. This relationship concerns, among other things, the introduction and the plurivocity of the term *interpretation*, according to whether it is used in a context of model theory or if it is taken in its common physical sense. In model theory, the excess of semantics (construction principles) with regard to syntax (proof principles) is first manifested in distinct interpretations. We mentioned the existence of non-isomorphic models with the same syntax, that is, non-categoricity (for example, the non-standard model of Peano's arithmetic, *cf.* above and Barreau and Harthong (1989)). Gödelian incompleteness furthermore demonstrates that some of these models realize different properties (elementary non-equivalence).

In physics, if we accept to see the equivalent of syntax in the mathematical structure of quantum mechanics and the equivalent of semantics in their proposed conceptual-theoretical interpretations (hidden-variable theories, for example, see below), then we would indeed find ourselves in a situation where the "semantics" also exceed "syntax" and, consequently, where a certain form of incompleteness (in this sense) would be manifest.[9] But more profoundly, in the case of quantum physics, it is the excess of the "possible" over the "actual" (that which is at play in the relationship between the characterization of the quantum state and the operation of measurement) which best illustrates this type of parallelism and of comparison. In other words, the result of a measurement corresponds to a

---

[9] If we want to continue with the analogy and in parallel with classical logic, we will also notice that this search for hidden variables – which prove to be non-local, however – evokes in a way the method of *forcing* which enables Cohen (1966) to demonstrate the independence in ZFC of the hypothesis of the continuum (by constructing a model that did not satisfy it, whereas Gödel had constructed a model that did). The hidden variables in question are indeed "forcing" within the physical theory they nevertheless continue to respect, in a somewhat analogous manner – conceptually speaking – to the one where forcing propositions are compatibly integrated with the original axiomatic model, in logic.

plurality of the potential states leading to it, each with its well defined probability. A sort of non-categoricity of the states (non-isomorphic, even if they belong to the same system) relative to the well defined result of the measurement, which operates here like a sort of "axiomatic" constraint in that it stems, as we have seen, from the universe of physical principles of proof (empirical proofs are "constrained" by physical measurment).

The concept of incompleteness was understood by EPR in the sense that quantum physics should be deterministic in its core and that its probabilistic manifestations would only be due to the lack of knowledge of "hidden variables" and of their behaviors, which actually amounts to saying "there are hidden causal relationships (not described by the theory) between particles." Now, the EPR argument is experimentally contradicted by the violation of the Bell inequalities (see above) and the property of non-separability is indeed inherent to quantum physics. In this sense quantum theory has been shown to be indeed complete.

Can we nevertheless speak of incompleteness in a sense different than that of EPR without, however, it being totally extraneous? In other words, would it be possible to formulate a proposition that is undecidable in the sense that it would be true according to one model and false according to another? Let's consider the crucial statement which can be attributed to EPR: "there are hidden variables." As we have just seen, this statement is false according to the usual model (standard interpretation) of quantum mechanics; yet a theory presenting the same properties as quantum mechanics can be constructed, in which this statement is true *on condition* that an adjective is added: "there are *non-local* hidden variables"[10]

---

[10]"Local variables" is an expression which is also equivalent to "variables attached to particles" (they depend only on properties specific to particles that happen to be local). To speak then of non-local variables is to express the fact that the variables which govern the behavior of particles are not only attached to these particles, but may involve others, which is another way to consider the non-separability we have just mentioned. As a matter of fact, the distinction between two particles having interacted is a representation that stems directly from classical physics – be if it isrelativistic. For its part, quantum physics proves to be fundamentally non-local. This is the reason why a type of true proposition in a model (inexistence of hidden variables and non-separability) can be false in another (existence of hidden variables, but by specifying *non-local* variables). In fact, more broadly, the controversy first initiated by de Broglie continues among some physicists (currently a small minority), with regards to the character of causal determination (classical but hidden, undescribed) of quantum physics.

2.2.1.4 *Incompleteness* vs *indetermination*.

Sometimes, confusion is set in between the concept of incompleteness (as presented by EPR) and that of *indetermination* (as highlighted by Heisenberg). So let's try to further explain the relationships between incompleteness and indetermination, which do not cover the same conceptual constructions.

The issue of incompleteness such as raised by EPR leads, we have seen, to the search for "hidden variables" which would "explain" the counter-intuitive behavior of quantum phenomena. It is indeed possible to elaborate hidden-variable theories, but these variables are themselves non-local, therefore simply postponing the intuitive difficulties. In this regard, it is better to preserve the canonical version of quantum physics, the Bell inequalities and the experiments by Aspect which highlight the property of quantum *non-separability* (that for two separated quanta, which have previously interacted, any measurement on one would determine the value of the other). By highlighting the fact that one of the origins of this situation stems from the *complex* number (state vectors or wave function) character of additive quantities (principle of superposition), whereas that which is measured is a *real* number and refers to the squares of the moduli of these quantities. We will return to this in a moment.

Quantum indetermination ("the uncertainty principle"), for its part, mobilizes somewhat different concepts: it consists in the treatment of explicit variables (non-hidden), such as positions and momenta, of which the associated operators will present a character of *non-commutativity*. In short, measuring first one observable, then the other does not commute: a fundamental property which will lead, moreover, to the development of non-commutative geometry, the current insight into the space of microphysics. It is in fact an issue of the constraints, which weigh the Plank constant (small but non-null) upon physical measurement, precluding simultaneous measurement with an "infinite" precision of two conjugated magnitudes such as positions and the corresponding momenta, which we have just mentioned.

If we wanted to roughly distinguish the two types of conceptual ambiguity introduced by these quantum properties with regard to habitual intuition, we would say that incompleteness refers rather to an ambiguity of object (Is the object local or global? How is it that according to the nature of the experiment it appears to manifest either as a particle or a wave?[11]), whereas indetermination leads to an ambiguity regarding what is

---

[11] Situations with regard to which Bohr was led very early on to introduce the concept of

92   MATHEMATICS AND THE NATURAL SCIENCES

meant by the "state" of the system. A quantum object of which we know the precise position would be affected by an imprecise velocity; conversely, the precise knowledge of a velocity would entail an "indeterminacy" regarding the position occupied – in a certain way, a quantum object which we could manage to "stop" would occupy all space.

In fact, both versions of these quantum specificities refer to a difficulty of describing quantum phenomena "classically" within time and space such as we experience them usually, while their description within their own spaces (a Hilbert space, for instance) is perfectly clear.

Regarding space, we will note several traits, which make of our usual intuition of space (and even of the Riemannian manifold of general relativity) an instrument which is unadapted to properly represent quantum phenomenality:[12]

i) Firstly, as we have indicated above, quantum quantities are defined at the onset on the field of complex numbers $\mathbb{C}$, as opposed to classical and relativistic quantities which are defined at the onset of the field of real numbers $\mathbb{R}$. It stems from this that in quantum physics, what is added (principle of superposition) is not what is measured (complex amplitudes are added, their square norms are measured). At the same time and for the same reason, quantum objects so defined (a wave function or state vector, for example) are no longer endowed with the "natural" order structure associated with real numbers.

ii) Then, as we have seen, the definition of the observables makes it so that some among them (corresponding to the conjugated magnitudes) are not commutable, as opposed to the classical case. In the context of the geometrization of physics, this necessitates the introduction of a geometry that is itself non-commutative, thus breaking with all previous traditions: as access and measure of space is non-commutative, in Connes (1990) geometry has been reconstructed accordingly, by using Heisenberg's non-commutative algebras as a starting point (since Klein and Gelfand, we know how to reconstruct geometry from – commutative – algebraic structures).

iii) Finally, the inquiry may lead even further, with regard to the rela-

---

complementarity (in the sense of a complementarity specific to the quantum object, susceptible to manifest, according to the type of measurement performed, either as particle or as a wave), which was the object of many controversies.

[12]For example, with inseparability, everything seems to occur as if an event locally well defined in a state space, that of the definition of quantum magnitudes, was to *potentially* project itself upon two distinct points in our usual state space.

tionships between quantum phenomenality and the nature of our usual geometric space: may the latter be Newtonian or Riemannian, it will admit a representation as a set of points and its continuum stems from an indefinite divisibility. Now given the existence of a scale of length (possibly minimal, *cf.* Nottale (1999)) such as the Planck length, recent string theories lead us to ask if in fact this space would not escape a description in terms of punctual elementarity (eventually to the benefit of another, in terms of interval elementarity,[13], or in terms of higher dimensionality such as "branes")

It should be clear that these specific issues lead to a conceptual revolution in our relating to physical space, at least the space of microphysics. The key idea is that geometry, as a human construction, is the consequence of the way we *access* space, possibly by *measurement*. So Euclid started by accessing, *measuring*, with rule and compass. Riemann analyzed, more generally, the *rigid body* (and characterized the spaces where this tool for measurement is preserved: those of constant curvature). Finally, today's non-commutative geometry, in Connes' approach, begins by reconstructing space by quantum measurement, which happens to be non-commutative (measure this and, later, that, is not equivalent to measuring that and, later, this, as we said). And this takes us very far from the space of senses or even classical/relativistic spaces. Yet, it may provide a geometric way to a novel unity, by explaining first how to pass from one mathematical organization/understanding of space to another.

2.2.2 *Finite/infinite in mathematics and physics*

2.2.2.1 *Some limits in physics*

A second essential theme emanates from the first part of this chapter: that of the fundamental role of *mathematical infinity* for both the constitution of mathematical objects as well as for proof. Mathematical infinity is often contrasted to a finitude constitutive of natural objects, in order to distinguish idealities and "real objects." It is indeed commonplace to hear that infinity has no place within the natural sciences, in physics particularly: all great objects which are the object of physics – including the universe itself – are they not finite? And as great as it may be, is the number of atoms in

---

[13] We would then maybe pass from a Cantorian representation of continuity to a representation by interval interlockings such as proposed by Veronese or to the nil-potent infinitesimals of Bell (1998).

the universe not likely to be finite?

Maybe such is indeed the case at the phenomenal level, but we would like to argue that in very essential cases in terms of *mathematical theory* and *modelization* of these phenomena (that is, also for the *construction of their objectivity*, their explanation, even their comprehension) it is quite different: actual infinity is present and even necessary, as so is continuity.

Of course, already at the level of differential equations, very present within physical theories, the use of the concept of derivative or of differential element already implicitly modelizes a representation of the "actualized" limit or of the infinitely small. And we know that there exist cases where discretization misses the specificity of the phenomenon (see Chapter 5). But it is not at this level that we want to base our argument, because in most cases, however, discretization and the use of large but finite numbers constitute a sufficient approximation to completely account for phenomenal manifestations to the degree of approximation of their measurement.

But there are also a number of physical cases where actual infinity plays an essential constitutive role in the theorization of phenomena. We will choose two which appear to be quite illustrative and which, moreover, present the interest concerning the modelization of the phenomena susceptible to immediate experience and that are readily available: the theory of critical phenomena (phase transitions) and the statistical theory of irreversibility.

The first case is that of critical phenomena. During a phase transition (as passing from a state to another), *non analytical* behaviors of thermodynamic magnitudes appear at the critical point. That is, critical exponents appears or the state function cannot be seen as the sum of a converging series or, also, some order of derivatives diverge. From a fundamental theoretical viewpoint, in the framework of statistical mechanics, this can only be accounted for by the passage to the actual infinite limit of the extensive magnitudes[14] of the system (N, the number of the system's constituents, V, the volume of the system), the corresponding intensive magnitude (d = N/V density) remaining constant. And here, we may find a first physical example of this finite/infinite correlation mentioned in the first part of this chapter, in a mathematical perspective. Indeed, the system's partition function Z[15] is generally an analytical function for any finite system in its

---

[14]The extensive magnitudes refer to an object's mass, size ...; intensive magnitudes have a character which is independent of "extension", such as a density, for example.

[15]In short, the partition function encodes how the probabilities are *partitioned* among the different microstates, based on their individual energies.

volume and in the number of its constituents. It is only at the passage to the actual infinity of this magnitude of the system that this function becomes non analytical and thus causes the critical behaviors to appear (divergence of correlation lengths or of susceptibilities, appearance of critical exponents). These traits are still heightened when the renormalization procedure is brought into play, "describing" in a way this transition, namely by introducing a recursive equation associated to the existence of a fixed point of scale transformations.[16]

The second case has similar traits, including the physical correlation of finite/infinite: we know that equations of particle movements interacting according to a potential (as within a gas, for example) are all reversible in time. So how does the irreversibility of thermodynamic properties appear? Beyond the statistical approach, the dynamic approach (dynamics of correlations) makes quantities appear, which, if they remain finite, maintain the reversibility in principle – the trajectories of the individual particles, say – but which, in the passage to actual infinity transform fractions with small denominators into "distributions." These are a form of generalized function, obtained as a limit, which allows us to describe divergence or discontinuities. This causes an irreversibility of phenomena, the passage to the limit (a mathematical analogy: from a limit, an irrational real number say, one cannot reconstruct – reverse – the specific converging sequence that led to it, see below). In short, the physical principles of thermodynamics, which lead to the crucial (an empirically evident) irreversibility of certain phenomena, may be understood in terms of particles' trajectories, provided that these are taken to infinity (the thermodynamic limit). Infinity is needed in order to understand the finite, or the irreversible behavior of a finite quantity of gas, say, is reduced to a simpler mechanics, if we understand it as the result of an infinite statistics.

Let's stress again that these physical situations, where the actual passage to infinity seems necessary to theorization and therefore to intelligi-

---

[16] *Renormalization* is a mathematical technique, in its modern methods, that is at the origin of quantum electrodynamics. From a conceptual viewpoint, it is a redefinition of the object in question, by adding to its initial characteristics (typically its mass) certain classes of interactions which modify them. More generally, the "renormalization group" provides the mathematical tool used to represent the passing from the local to the global along the critical transition, in many domains of physics: renormalization describes a change in measurement and of object, obtained by integrating the new classes of integration due to transition. Its properties are due to the fact that to critical transition, the passage to the infinite limit of the correlation length entails an invariance of scale of the system and brings forth a fixed point from the dynamic.

bility, present conceptual traits shared with certain purely mathematical situations. Thus, for any finite n, as great as it may be, the sum of the sequence $1, \ldots 1/n!$ belongs to the field $\mathbb{Q}$ of rational numbers. It is only for the value n at the "actual" infinite limit, $n = \infty$, that this sum strictly "exits" this field to provide the transcendental number e belonging to the field $\mathbb{R}$ of real numbers without belonging to $\mathbb{Q}$.

Likewise, distributions (such as Dirac's δ function) are considered as "exiting" the function domain by the passage to actual infinity of sequences of functions (like the Gaussian or Lorentzian functions, in this specific case).

These situations (and many others) appear to be quite related, conceptually speaking, to those we have described in the preceding physical cases. While in most computations finite approximations of non-algebraic numbers may suffice, the construction of the reals, as well as the mathematical understanding of critical phase transitions or thermodynamic irreversibility require the actual, infinite limit. In mathematical physics, thanks to the actual passage to infinity, it is always an issue of constituting new "objects" (or a new phenomenality), which otherwise would be theoretically inaccessible. It is a way of going beyond a given objective or conceptual frame and constructing a new one.

In a similar line of thinking (although objective phenomenality is obviously quite different), we know that in the quantum theory of the hydrogen atom, the electron's energetic states present two types of spectra: a discrete (countable) spectrum while the electron remains linked to the proton, and a continuous spectrum after the liberation of this electron. The threshold between the discrete and continuous (mathematical) characteristics is associated with a radical change in the electronic physical state and to an exit of the initial domain of characterization: the *hydrogen atom's* electron, exiting its domain of definition, becomes a *free* electron and moves into a new physical frame.

2.2.2.2 *Constitutions of objectivity. Conceptual comparisons between mathematics and physics*

Everyone agrees to say that there does exist, among humans, a sort of first intuition of the topological continuity, associated among other things with tactile experience (*cf.* Desanti for example) or to motor experience (*cf.* Thom or Berthoz). Likewise, "actual infinity," once conceptually constructed by constitutive intersubjectivity, seems henceforth to stem also from the subject's experience: this experience is not only conducive to imagining the idea of God (and theology) or, like Bruno, the infinity of worlds. It is, even more effectively, involved in the process of constitution of mathematical objects (both finitary and infinitary) in that it may be considered that mathematical objects are "substantially," if we may say so, a concentrate of actual infinity brought into play by human beings. Let's explain and argue.

The transcendental constitution of mathematical objectivities, from the completely finitary objects such as the triangle or the circle, to the structures of well-order, or even to the categories of finite objects of which we have spoken, actually involves a very fundamental *change of level*: the process of abstraction of acts of experience and of the associated gestures (see part 2.1) leads indeed to this *transition* which constitutes the forms thus produced into abstract structures, into eidetic objects. This transition, which also leads forms to their conditions of possibility, presents all the characteristics of criticality which may currently be considered and, notably, this passage from the *local* (such or such empirical form) to the *global* (the structure which is defined abstractly and which is to be found in any particular manifestation), as well as the passage from a certain ("subjectivizing") heteronomy to an autonomy (objectivity), and from a certain instability (circumstantial) to a stability (a-evential). It is namely these characteristics, resulting from a sort of passage to the effective infinite limit as a process of constitution, which probably correspond to mathematical platonic thinking, which, however, forgets the constitutive process itself. The Greek invention of the abstract line with no thickness is an early paradigm of this kind of "critical" conceptual transition. It raises the question whether also the construction of a triangle, the mathematical one, shouldn't also be considered as based on an implicit use of actual infinity: go to the physically inexisting limit of lines with no thickness.

More rigorously, the images we can use to try to account for this transition, which really leads to the constitution of these new objectivitie are varied. In mathematics, we have used earlier the image of the sum of rational numbers ($1/n!$, for example) which lead in certain cases to the actual infinite limit, on a transcendental number (e, in this case), a true critical transition which leads from the exit of one field ($\mathbb{Q}$) to enter into another one encompassing it ($\mathbb{R}$). In physics, we may find something like an equivalent in a change of phase, associated, as we have said, with the divergence of an intensive magnitude of the system (a susceptibility, for example) and to the passage from local to global (divergence of the correlation length of interactions). In biology, it would be a case of a change in the level of functional integration and regulation (the organism in relation to its constituents, for example).

If mathematical structures are also the result of the search for the most stable invariants, as is conceptually characterized in the preceding, it is then probably due, at least in part, to this process of constitution mobilizing a form of actual infinity leading to a sort of stable autonomy.

Let's continue with the physical metaphor of phase changes. A phase transition can be manifested for instance in a symmetry breaking of the system and a concomitant change, sudden or more progressive, of an order parameter (the total magnetic moment for a para-ferromagnetic transition, density for a liquid/solid transition). In fact, the phase transition is, in one way or another, a transition between disorder (relative) and order (also relative). If we keep these characteristics in mind and make them into a conceptual trait that is common to the transcendental constitution of mathematical objectivities, we will readily notice the disordered situation with regard to the often uncoordinated collection of "empirical" mathematical beings, and the ordered situation in mathematical objects and structures as such, as resulting from the process of abstraction and of constitution. So it is easily conceivable why the axiomatization, or even the logicization of the statements characterizing these mathematical structures, are genetically and in some respects conceptually *second* relative to the mathematical activity and to the process of constitution itself, as we emphasize in the first part of this chapter. Indeed, these statements describe in fact *order* consecutively to the "phase transition" we have just invoked. But mathematical thinking concerns itself as much with the process of transition as the putting into form and description of its result. And to go even further, we could almost say that the evacuation of the "disorder" accompanying transition-constitution corresponds to the evacuation of the "significations" associated with the structures over the course of their elaboration and to the "infinitary involvement" it presupposes. This it what probably enables "structuralists" (in the sense of Patras (2001)) to assert that the return to foundations, such as conceived in logical-formal thought, leads to pure syntaxes devoid of meaning.

It is probably also one of the reasons enabling us to understand why formalism "works" when it is an issue of describing the order resulting from the transition in question (the constituted mathematical structures), but that it fails from the moment it is given the task of *also* describing the transition itself, that is, the process of constitution as such, from and in the terms of its result. In fact, one may consider that formalism fails to capture "actual infinity" that enabled the passage and which has become a major characteristic of the objectivities thus constructed,[17] and one may wonder if it is not natural to also interpret the results of incompleteness as a true demonstration of this lack.

So here ends, in our view, the proposed conceptual analogy with phase transitions in physics because we know that in physics, as we have recalled earlier, renormalization theory proves itself, in a way, to be able to address the critical transition itself (see the last note in sect. 2.2.2.1). This difference in behavior relative to the processes put into play is doubtlessly to be referred to the difference between the objects considered themselves, such as they are elaborated in physics and in mathematics: if the construction principles are similar, as we have seen, on the other hand the proof principles are completely different, and it is indeed regarding the status of the proof that the difference is manifest.

It is probably what transpires in this dichotomy introduced in the first part relatively to elementarity, by opposing simple elementarity (related to the artificial processes of algorithmic calculus, to the concatenation of simple logical gates, or even to any artifact) and complex elementarity (related to natural processes) such as strings in quantum physics or cells in biology. Indeed, inasmuch as physics (and biology) address natural phenomena, they are only likely to be confronted with elementarities that are rather complex and hence with ones that are truly irreducible to simple elementarities in the sense of artificial computation. It is also probably that which prevents us from reducing scientific judgment to a calculus (in this sense), without however denying the interest of the complementary understanding conceptually provided by the computational attempt.

To conclude That the action and the gesture precede any comprehension, or if it is presented as a condition of possibility for this comprehension, is something that biology has taught us even prior to studies on cognition, particularly evolutionary biology. It appears that neurons, nervous systems, and cerebral structures differentiate themselves and develop only among organisms endowed with motor autonomy. It is therefore not sur-

---

[17]In this sense that it is likely that it is the so-called "semantic" aspect, linked to significations, that is the most deeply involved in the occurrence of this effective passage to the infinite limit, whereas the syntactic aspect is much more relative to the rigorous description of the results of this passage. It is indeed the non-coincidence of these two dimensions that is at the origin of the properties of incompleteness or of non categoricity, or even of the discrepancy between proof principles (formal) and construction principles (conceptual), of which we speak in part 2.1.

prising that the most general and most abstract concepts, in mathematics and in physics, are "firstly" rooted in motor experience and that it is "the gesture" which enables its development at the same time that it enables, as abstract as it may have become itself, the elaboration of new concepts.

Three gestures seem to have presided the birth of the most fundamental of mathematical and physical concepts (see also Salanskis, 1991): that of "displacements" for *space* (and time), that of the caress for *continuity*, that of unlimited iteration for *infinity* (and *numbers*). The passage to the limit of these notions, which are firstly tributary of their intuitive grasp correlative to action, confers them their autonomy of rigorous and defined concepts: space and time are no longer reduced to being receptacles for our movement, continuity no longer gives rise to insurmountable paradoxes (eleatics, infinitesimals), numbers no longer designate only integers, but also real numbers, and infinity itself is apprehended and becomes actual. And what the abstraction of movement constitutes from what is authorized by what the gesture makes, with mathematical physics, a theory of physical matter, possible and fecund.

Chapter 3

Space and Time from Physics to Biology

Premise This chapter offers a twofold epistemological analysis of the concepts of space and time: the first section frames them in the setting of contemporary physics, the second section deals with their role in biology and especially in the project of its mathematisation. Both investigations are closely connected with questions in cognitive science. The issues involved in the analysis of the foundations of mathematics and the natural sciences have profoundly affected approaches to human cognition and the treatment of these foundational questions forms an indispensable preliminary to our understanding of life and cognition.

Contemporary physical theories have led to a steadily more pronounced geometrization of physics, the counterpart of which has been a steadily more pronounced physicalization of geometry. This is clearly illustrated in general relativity, where the geometrization of gravitation (the trajectories of objects are described as geodesic curves – or optimal paths – in a Riemannian space manifold, see Chapter 4) can equally well be interpreted as the physical realization of a mathematical structure (the space-time curvature is determined by the distribution of energy-momentum). This geometrization is seen even more clearly in quantum field theory, where the introduction of non-commutative gauge fields[1] to give an account of the dynamics of interacting fields has led to the development of an intrinsically non-commutative geometry (see Connes, 1995).

As for the epistemological status of space-time concepts, the mathemat-

---

[1] Relativity theory, in particular by the work of Weyl, introduced gauge theories as field theories in which the Lagrangian (the total energy) is invariant under a certain continuous group of transformations. These fields are at the core of general relativity, as they describe the invariants under change of reference systems. The non-commutative case is meant to deal with non-commutative measurements in quantum physics, as formalized by Heisenberg's non-commutative algebra of matrices.

ical specification of geometric notions can be seen as a process of the objectivization of the forms of intuition of our phenomenal awareness. Indeed these very forms of intuition, just as much as the mathematical specification of the structures of space and time, are to be investigated within the setting of specific contemporary physical theories. When we turn to the role of mathematics in biology, the *constitutive* role which mathematical concepts play in physics is in contrast to their prevailing *conceptual* status in biology. The various affordances and regularities which experience furnishes are transformed in physics into very rich mathematical structures – structures far richer than suggested by the "symptoms" through which our senses and/or physical instruments view the physical world. Moreover, these mathematical concepts, rather than being merely descriptive, play a regulative role in constituting our concept of physical reality. One can say nothing of the subject matter of relativity, of quantum theory, or of the general theory of dynamical systems (the heart of theories of critical states and phase transitions) without mathematics.

In biology, by way of contrast, one is struck by the enormous richness of structure with which living systems, as given to us in phenomenal awareness, are already endowed, while their theoretical formulation in terms of mathematical concepts suffices to model only certain aspects of that structure. And this is so in a manner which tends to fragment their organic unity and individuality and fails to do justice to their immersion in wider ecosystems. If we reflect on the role of mathematics in human cognition we are thereby led to re-examine its role in biology, since an analysis of living systems are the starting point of any reflection on cognition.

Nevertheless, despite these differences and granted the lesser extent of overall mathematization in biology, one can recognize in many areas of biological research an apparent movement towards what may loosely be termed "geometrization."

Questions involving our understanding of spatial concepts are posed not only in the study of macromolecular structures (e.g. the organization in space of DNA base pairs and the resulting expression of genetic effects, or the investigation of the spatial structures of proteins or prions) but also within developmental biology (in the study of the effects induced by spatial contiguity in embryogenesis for example). The same may be noticed in the study of organic function (e.g. the fractal geometries affecting the boundaries of the membrane surfaces engaged in the regulation of physiological functions) and in the study of population dynamics and its associated environmental context. Alongside these areas involving spatial understanding, the examination of temporal concepts is also strongly implicated in such areas as the study of the response times to external stimuli, the iterative character of internal biorhythms, and in the study of synchronic and heterochronic patterns in evolutionary biology, the outcome of which has been a recent formulation of the synthetic theory of evolution itself.

What connection can we trace between the roles of spatial concepts in physics and in the life sciences? The conceptual scaffolding of modern geometry is itself rooted in the conditions of possible actions and experiences, which are a basic aspect of our presence in the world. It has at its foundation an inseparable intertwining between (i) our presence in the world as sentient creatures and centers of inter-subjective awareness, through symbolization and abstraction, and (ii) the evolutionary leap to which this capacity for rational thought and creative imagination has led. Such a constitutive *braid* connects the phylogenesis of humans to their ontogenesis as cultural beings in history, via the stabilization of inter-subjectivity through language. In this perspective we should also view the semiogenesis of conceptual constructions that arise in mathematics and physics. Without the initial spatiality of actions (especially gestures, with their intentionality) and the dimensionality of our primal imagination and cognition, we could never have arrived at the idea of a 10-dimensional manifold, in terms of which the theory of superstrings in quantum physics is elaborated.

In the first section we analyze the notions of space and time as characterized by three types of physical theories: relativity, quantum theory, and the theory of dynamical systems. In sections 2 and 3, we examine the same notions in connection to theoretical biology. We conclude by putting forward a tentative categorization, in abstract conceptual-mathematical form, of the manner in which space and time operate as invariants in determining our forms of knowledge.

3.1 An Introduction to the Space and Time of Modern Physics

3.1.1 *Taking leave of Laplace*

The physics of the nineteenth century carries the imprint of Laplace. His achievements in mathematics, physics, and philosophy marked the moment at which the development in the direction of modern physics, initiated by Galileo, Descartes, and Newton, reached maturity. Laplacian mechanics is organized in the framework of an absolute background space with Cartesian co-ordinates in which the motion of bodies is governed by the laws of Galileo and Newton. The perfection of this *mechanica universalis* is completely expressed through eternal mathematical laws. To cite Laplace: "An omniscient being with perfect knowledge of the state of the world at a given instant could predict its entire future evolution with perfect precision." But what counts for even more in Laplace's work, for us earthbound and imperfect beings, is not this divine, and unachievable, knowledge but the approximate analysis of (possibly perturbed) systems. If one knows the state of a physical system to a certain degree of approximation, one can in general determine its evolution to an approximation of the same order of magnitude.

In this sense, according to Laplace, mathematics rules the world and permits the prediction of its future state, by a finite and complete system of differential equations. In fact the analysis of the perturbations of planetary orbits was one of the chief impulses driving the development of nineteenth century mathematical physics. As for causality, in Laplace's approach, explicit determination (typically, by a system of differential equation) implies predictability.

The development of twentieth century physics has taken a quite different direction. Relativity, quantum theory, and general dynamical systems have led to an entirely distinct set of concepts and inspired a quite different philosophy of science from that which prevailed in the nineteenth, in particular, for causality.

We cannot say the same of the mainstream in the cognitive sciences. Turing, in his seminal article founding the strong AI program and setting out the functionalist account of cognition, made the explicit hypothesis underlying his generalized discrete-state machine (the "Turing machine"): "by its discrete nature, full predictability is possible, in the sense of Laplacian determinism" (Turing, 1950). The Laplacian idea of a finite and complete set of rules is thus consciously placed at the heart of the "imitation game" (envisaged in Turing's paper) through which he set out to demonstrate that the functioning of the brain was equivalent to that of a Turing machine.[2]

---

[2]See the reference to Laplace above and, by contrast, the example of "the displacement of a single electron which could lead to a man being killed in an avalanche a year later or to his escaping" in Turing (1950). Turing explicitly observes that "the nervous system is certainly not a discrete-state machine. A small error in the information about the size of a nervous impulse impinging on a neuron, may make a large difference to the size of the outgoing impulse." Yet, he believes that, if the interface is limited to a teleprinter, one should not be able to distinguish a machine from a man (or a woman?). Unfortunately, Turing was not aware that subsequent results on the geometry of dynamical

In fact the notion of a deterministic program, as it emerged in the work of the logicians of the 1930s (the theory of computability was developed by Curry, Herbrand, Gödel, Church, Turing, Kleene, and others in the years 1930-36) is inherently Laplacian, as clearly spelled out by Turing. That is to say, it implies complete predictability of the states of a computer running a program. More precisely, for Turing, unpredictability is just "practical," not by principle; it may depend, for example, on the huge length of a program. From this ideal model, which stems from the logical calculi of formal deduction rather than from physics, the Laplacian paradigm of the brain as a Turing machine running a program has been transferred to the study of cognition in the biological setting. It is of crucial relevance to the project we are pursuing here that the abstract description of a Turing machine is in no way dependent on our understanding of it as a spatial structure. The "Cartesian" dimension of its material being has no influence whatever on its expressive powers, as we shall see. Moreover, its internal clock records a sequence of discrete states in an absolute Newtonian time. It was explicitly invented and behaves as a logical machine, not a physical mechanism (see Turing, 1950; Longo 2007). By contrast, the analysis of space and time and their dimensionality is at the heart of any analysis of physical phenomena.

In relation to any claim that living systems and their mental activities can be "reduced" to physics we ought to ask: to *which* physical theory? *Which* physical laws have to be employed in the analysis of biological and cognitive phenomena? Functionalism is the still prevailing approach to cognition and biology (the "genome is a program" paradigm, for example) and implicitly refers to a Laplacian causal regime.

3.1.2 *Three types of physical theory: Relativity, quantum physics, and the theory of critical transitions in dynamical systems*

Relativity. Relativistic theories introduce a four-dimensional space-time in which conservation laws and relativistic causal principles are described in terms of invariants with respect to the relativity group of the theory. In special relativity (SR), the objects of the theory and the space-time structure are given together with their invariance properties under the group of rotations and translations in this space (the Lorentz-Poincaré group). In

---

systems would have confirmed the early work of Poincaré. In particular, no finite grid of inspection can stabilize a (possibly) unstable dynamics (see Longo, 2007, for details).

general relativity (GR), physical space is described as a Riemannian manifold of all possible locations together with its dimensionality and symmetry properties. The metric coefficients *are* the gravitational potentials just as the local curvature of the Riemannian manifold *is* the energy-momentum. Thus geometry *constitutes* the invariants we name as "objects" and "physical laws." It is not just that physical concepts acquire meaning within the framework of a mathematical space – the latter actually prescribes a thoroughly structuralized notion of *objecthood* and *objectivity* as invariants of geometrical structures.

In metric spaces, which carry the record of and themselves serve to record the cohesion of and between objects (the stability of physical laws and the conservation of energy and momentum), symmetries and geodesics shape the physical content of the theory. As we will extensively discuss in Chapter 4, Noether's theorem (1920s) describes these physical invariants in terms of space-time symmetry groups. Energy conservation for example is closely tied to invariance under the symmetry group of temporal translations, just as the geodesic curves furnish the trajectories along which quantities are conserved (inasmuch as they are stable minimal paths, see Chapter 4).

The underlying unity of SR (which unifies electricity and magnetism) and GR (which unifies gravitation and inertia) is reflected in the fact that SR may be considered a particular limit of GR. Once again we see geometry providing the framework for actually constituting new structural invariants and unifying them in the same space inasmuch as the stable properties of physical systems with that structure arise in connection with new groups of spatiotemporal transformations.

But there is also another path in the direction of increasing mathematical abstraction: the generative role of mathematical ideas provides the basis for grasping the sense of new physical concepts, indeed constitutes it. Take, for example, the physical applications one can find for the compactified (numerical) real line: one takes the infinite real line, an extra dimension, and transforms it into a circle, by adding one point (which "represents" infinity). On that basis one passes from four dimensions (three of space plus one of time) to five, but this fifth dimension is derived mathematically from the Lagrangian action associated with a field which is both electromagnetic (hence classical, i.e. non-quantum) and gravitational: this is the core idea for unification in SR, concerning the Maxwell and Einstein equations (for electromagnetism and gravitation, respectively). The physical properties carried by this new dimension of space are thus obtained by a mathematical extension of physical space – the fifth dimension, which is compactified, that is, it is folded over on itself in the form of a circle (Kaluza–Klein theory[3]). The geodesic principles and the symmetries are conserved. The observables of the theory have not changed, because the fifth dimension of this spatial structure is not a physical observable – it is a pure consequence of the conceptually generative capacity of the mathematical formalism.

At the same time this new dimension contributes to the *explanatory* power, for it allows us to unify the structure of theories which were formerly quite distinct, while exactly preserving the invariants (energy-momentum, etc.), which were at the heart of the two approaches. It is mathematical geometry, which provides us with this new physically intelligible space; and, through this geometrization of physics, mathematics plays a role of extraordinary explanatory power. In fact it supplies the models of space and time furnishing *the* framework for physical phenomena and gives them meaning. The required mathematical ideas are not laid up in advance in a Platonic heaven, but are rather constituted within the interface between ourselves and the world which they serve to organize conceptually. Recall the role of Riemannian geometry in organizing the framework of relativistic physics. Relativity indeed furnishes one of the most beautiful examples of this *mathematical constitution of phenomena*: the most stable and coherent part of our conceptual apparatus – mathematics - provides the framework for a structuralized conception of space and time which undergoes reciprocal adjustment as it encounters that source of friction (the world). This friction is continually suggesting/imposing new regularities to be incorporated in the structure, by canalizing our active conceptual construction toward some models or deflecting it from others.

Quantum physics. Relativistic theories present space-time as external to physical objects, aiming to understand the latter as singularities of a field, and their evolution as controlled by geodesics. In this case, their phenomenal appearance amounts to nothing more than the mathematical stability of the invariants attached to these geodesics. Quantum mechanics on the other hand adds to this external frame of reference (Minkowski space) an internal frame of reference expressed in terms of quantum amplitudes and their invariants. This internal frame of reference is essential because

---

[3]This theory was proposed in the 1920s in order to unify electromagnetism and gravitation. They introduced, within a non-quantum framework, a five–dimensional manifold, of which the fifth dimension, intimately related to the system's action, was compactified, thus generating a topological space–time manifold of $R^4 \times S^1$, not R^5. This Kaluza–Klein method served in a way as a foundational paradigm also for ulterior unifications (see Lichnerowicz, 1955, and Duff, 1994).

the atomicity implicit in quantum theory is a matter not, as in classical atomism, of smallest possible bodies in space, but rather an atomicity of the processes determining the evolution of the field (because the dimensions of the Planck constant are that of an action, i.e., energy multiplied by time). It is thus the variation of energy in time which is discretized in quantum theory and not the structure of matter or of space-time.

Space and time remain continuous, as in relativity,[4] and this remains true, in certain respects, of quantum fields, although they behave in a different manner from classical fields. However, the mathematical unification of the theory of quantum fields with that of the gravitational field is far from being accomplished.

Our understanding of global or external spaces is profoundly bound up with that of local or internal ones: particles, as much as fields, display counter-intuitive non-local effects. In short, quantons or quanta[5] can be present simultaneously at widely separated locations. This behavior is not magic: matter fields are not local – they are not reducible to space-time singularities as in GR. Furthermore, matter includes fermionic fields. On this point, debate centers on the relationship between internal and external spaces – and the debate is very lively, notwithstanding the Einstein–Podolsky–Rosen paradox which had appeared to demonstrate the opposition between GR and the physics of quanta. Briefly, quantum mechanics, which in first approximation had appeared to bring no essential new element to the determination of our theoretical notions of external space, has nevertheless introduced a novel (and counter-intuitive) perspective on our notion of locality. On the one hand, the physical laws of quantum mechanics remain local in the sense that the evolution of a system between measurements is generated by partial differential equations. On the other hand, the characteristics of the probability amplitudes associated with the state vectors (complex numbers, the superposition principle) engender a non-separability in the properties of quantum systems which is bound up with

---

[4] In mathematical terms, the external space-time constitutes the base space of a fiber space, the fibers of which (derived through generalizing the notion of the inverse of a Cartesian projection) serve to organize the structure of the internal spaces. But the external space-time of quantum physics, considered as the base space of a family of fibers, displays in general a continuous topology corresponding to the classical representation of special relativity. Discrete processes – such as the quantisation of energy or spin – involve these additional, internal dimensions, along the fibers.

[5] The term "quanton" designates a quantum object which is susceptible to manifestation in either its particle or wave aspects depending on the experimental set-up (metaphorically: according to what question is put to the system).

measurement and corroborated by experiment (Bell inequalities and the Aspect experiment concerning quanta which have interacted in the past[6]).

Despite the absence of theoretical unification, there are mathematical invariants which carry over from the local to the global frame of reference and vice versa. For instance a global shift in the reference system does not alter the electric charge: certain measurements are locally and globally invariant (in the theory of gauge fields) and the fields themselves are associated with local gauge invariants. Super symmetric theories best tend to illustrate the connection between internal and external spaces. In these theories one can adjoin further dimensions to the four of relativistic space-time, in the manner of the Kaluza–Klein compactification of space, with the aim of preserving, as far as possible, the space-time symmetries; recent theories of quantum cosmology have sought to unify the theories mentioned here, in a tentative yet very audacious manner, at the level of the Big Bang by representing space as a six-dimensional manifold in which four dimensions would expand (the four-dimensions of the observed universe) while the compactification of the other dimensions provides for the way in which the properties of matter (fermionic fields) and interactions (bosonic fields) are structured. We should also mention the possible role of the non-commutativity of quantum measurements (the complementarity of position and momentum): a fundamental difference from classical and relativistic approaches. A very promising framework for unification has been proposed via a geometric approach (by Connes, in particular). The idea at bottom consists in reconstructing topology and differential geometry by introducing a non-commutative algebra of measurements (the Heisenberg algebra) in place of the usual commutative algebras, see Connes (1995). Once again, the geometric (re-)construction of space has the effect of making (quantum) phenomena intelligible.

Dynamical systems and their critical behavior. The physical theories of the type we next consider are concerned with dynamical systems, which, for some values of the control parameters (e.g., temperature), display discontinuous or divergent evolution (phase transitions such as the freezing of liquids), progressive transition from ordered to disordered states (as in paramagnetism and ferromagnetism) and qualitative change in their dynamical regimes (such as bifurcations of phase-space trajectories or transitions from cyclic to chaotic behavior). They may be regarded collectively as theories of

---

[6] Aspects of this nature have nourished more holistic conceptions such as the ideas of David Bohm, Basil Hiley, and their collaborators concerning the so-called "implicate order."

phase transitions. In approaching the question of causality by the status of space and time in these theories, we must distinguish between two classes.

The first class of theories concern systems which possess a high number of degrees of freedom: the phase space is therefore very large, as in thermodynamics and statistical mechanics. It was in relation to this class of theories that problems relating to temporal reversibility and irreversibility were first posed. The second class of theories is concerned with nonlinear dynamical systems which can be treated only in terms of a small number of degrees of freedom, and the properties of whose dynamics (bifurcations, existence of strange attractors, etc.) are associated precisely with the nonlinearity, whether treated within the framework of continuous differential equations or via discrete iterative procedures. These systems also pose questions of reversible or irreversible behavior, but in slightly different terms from those in the first class. In both cases, and in contrast with the situation prevailing in relativistic and quantum theories, where we find ourselves in a fairly regular universe, here our attention is more on the *singularities* than the *regularities* of the systems in question.

Both these classes of theories mark an apparent return to more classical conceptions of space and time than those encountered in connection with relativity or quantum theory. In particular, the introduction of spaces with a large number of dimensions (such as the phase space of statistical mechanics) does not involve their fulfilling the sort of constitutive role assigned to space-time structures in relativity or quantum theory. Nevertheless, these two classes of theories have also given rise to new approaches to physics, this time relating to aspects which are, on the one hand, in relation to space, markedly morphological and global; and on the other hand, in relation to time, markedly evolving and directional; and this marks the causal relations.

Yet, these systems are characterized by numerous other traits. One is the role they frequently assign to the global aspects. If one takes the "most simple" dynamical system, three bodies together with their associated gravitational fields, the very unity of the system prevents its being analyzed in Laplacian terms. One cannot know/predict the position and momentum of each body without at the same time analyzing the same parameters for the others. They are correlated through their mutual gravitational fields so as to constitute a sort of *holon*: a global configuration, which, evolving in time, determines the behavior of each of its elements. A step-by-step analysis – that is to say an analysis of the behavior first of one body, then two, or the approximation of that behavior via Fourier analysis –

is simply not possible here. This is what robs the system of the kind of completely predictable behavior conceived by Laplace. What wrecks this Laplacian predictability is that in sufficiently complex dynamical systems (in the three-body problem rather than the two) divergences are present (i.e., discontinuities related to control parameters). The non-linearity of the mathematical representation reflects the intrinsic unity of such systems. The dramatic change, as for knowledge and causal regime, is due to the fact that determination, under a finite set of equations or inference rules, does not imply predictability.

Note that dynamical systems are often assigned their *proper* time in a "peremptory" fashion. Insofar as they exhibit phase transitions, by the bifurcations, as well as their transitions from cyclic to chaotic regimes, these "impose" directionality on the states of the system, differently from other physical theories. That is, their time is oriented by phase transitions and, irreversibly, by bifurcations and transitions to chaotic behavior. The essentially irreversible character of time for these systems marks a definite contrast with the picture of time in relativity, where it is intrinsically reversible and its flow is regarded as an epiphenomenon. The irreversibility of time characteristic of such "critical" systems is connected with their unpredictability and their chaotic behavior. Note finally that this seems to provide a concept of time more appropriate to living beings, which are strongly affected by thermodynamic phenomena amongst others, by the role of energy production in all living organisms.

3.1.3 *Some epistemological remarks*

We have examined aspects of the geometrization of modern physics. The mathematics of space and time molds a framework for the understanding and organization of phenomena and the unification of different "levels" of their structure. The epistemological and mathematical aspects of space and time turn out to be profoundly bound up with one another in a manner which plays a pivotal role in shaping scientific inquiry, in particular in providing for the unification of physical theories. We have briefly mentioned the (pre-quantum) unification of electromagnetism (governed by the Lorentz-Poincaré group) and gravitation (governed by the group of diffeomorphisms of GR). More recent theories introduce new symmetries (supersymmetries or symmetries of the space-time structure in a generalized sense, associated with the notion of super-space) allowing the articulation within a common framework of the external and internal spaces of quantum systems. From an epistemological standpoint, the unifying aspect of these theories is that they lead to the construction of unfamiliar spaces whose physical relevance is then corroborated by experimental investigation.

More recently still, a non-commutative geometry has been invented, in reference to the non-commutativity of quantum measurements, and we have, hence, been led to propose geometric structures even further removed from the ones directly suggested by the world of senses. Geometry provides a mathematical framework organizing the practical as well as the theoretical aspects of our spatial experience. Our access to space as expressed in the most developed physical theories is based on measuring instruments very far removed from our naive sensations and, hence, necessarily follows a route for the (re-)construction of our notion of space very different from what these might suggest. The curvature of light is detectable only through sophisticated astrophysical measurements, it is not apparent to our raw intuition. The geometry of the universe rests on a geodesic structure quite unfamiliar from the viewpoint of sensory experience. The non-locality of quantum phenomena follows from microphysical measurements quite inaccessible at the level of our physiology.

It is even possible that our geometry itself will take the form of mathematical structures in which the classical notion of a point is no longer basic (e.g., the theory of superstrings or twistor theory). Notice besides this that the generalization – via homotheties – to all physical scales and dimensions, of all Euclidean properties drawn from our sensory experience is quite arbitrary. Straight lines and dimensionless points do not exist, or "exist" in only the same sense as any other mathematical construct or abstraction. They can be replaced by other abstractions which may turn out to hang together better with experimental evidence and with new tools of measurement.

One last word about theories of dynamical systems, near to or undergoing critical change. The treatment of space-time suggested by those theories (centered on phase space and the transition of their dynamics from stable to chaotic behavior), introduces new elements important also for other theories, above all in connection with certain recent cosmological theories (models of the phase transition associated with the Big Bang, singularities, and cosmic strings, for example) and also in connection with the relations between local and global structure.

This class of theories forms a key bridge between physics and the life sciences. Moreover, we must skate fearlessly over a great many intermediate levels of organization, with their major associated critical "epiphenomena" – namely, cognitive phenomena.

3.2 Towards Biology: Space and Time in the "Field" of Living Systems

3.2.1 *The time of life*

As a preliminary, we want to analyze the particular features of time characteristic of living systems. Temporal *irreversibility* is at the heart of the study of dynamical systems exhibiting critical behavior, but it is also characteristic of living systems. At every stage phylogenesis and ontogenesis are marked by "bifurcations" and by the emergence of unpredictable phenomena and structures, which resemble those observed in critically sensitive dynamical systems. Moreover, living systems contain a great many subsystems, which display this kind of critically sensitive behavior – dynamical and thermodynamical. These contribute not only to the temporal irreversibility of the system but also to a kind of unity, which is apparent in the kind of dynamical systems we touched on above in connection with the three-body problem. Poincaré's three bodies, in exhibiting an example of this kind of *unity*, form a primitive *gestalt* associated with a purely gravitational interaction. Two bodies exhibit a quite different dynamical behavior, stable and predictable. It could even be said that what comes into play in the three-body dynamical regime is a kind of emergent behavior, a unity of non-stratifiable relationships: one cannot analyze first the position, then the velocity, of each body step by step, independent of the unity of the system they form.

In the light of what we said, we may summarize our perspective as follows: the spatial and temporal terms do not appear to possess the same significance or play the same role within the two principal approaches: "geometric" *vs* "algebraic-formal." In the "geometric" approach, space is the correlate of geometry itself, it intervenes at the perceptual level. Time is the time of genesis of structures, the recording medium of their process of constitution. In the "algebraic-formal" approach, by contrast, spatiality is the echo of an abstract linguistic inscription, of formal descriptions, while temporality seems to be principally a matter of sequential functioning, of the execution of algorithmic calculations.

This remark refers to the distinction, which we have drawn above in the context of the foundations of mathematics, between *principles of construction* (in particular those with a geometrical aspect) and *principles of proof* (formal principles of logic). Mathematics is built up on the basis of both types of principles. The philosophical fixation, implicit in the analytic tradition, with logicism and formalism has tended to exclude or sideline the first of these. The "linguistic turn" has given us extraordinarily rich logical/formal machinery (and literally machinery in the form of digital computers) but it has also endorsed the myth of the complete mechanization of mathematics, indeed of any form of knowledge.

We have argued that incompleteness theorems in formal systems are due to the difference in expressive strength between these two types of principles. Conceptual constructions based on space-time regularities possess an autonomy, an essential independence in relation to purely formal descriptions, the latter being made exact by mathematical logic, through the work of Hilbert and his school. Unfortunately physicists are prone to label any "mathematical treatment" of a subject as an instance of "formalization." For logicians these are quite distinct notions: there is the Gödelian (and other forms of) incompleteness in between. The distinction of principles of geometric construction from algebraic-formal principles of proof is in our view one of the crucial factors which underlies the constitutive role of space-time concepts and geometry in the analysis of cognition.

In the conception of time as the medium of algebraic manipulation and formal calculation, as seen in the sequential running of a computer program, one recognizes an important fruit of the formalist view of the foundations of mathematics. The 1930s marriage of Hilbertian formalism, together with the problems it addressed (the completeness and decidability of formal systems) and, on the other hand, a mechanistic positivism, was at the origin of the attempt to treat human rationality in terms of a mechanism, which indefatigably executes formal algorithms.

But this forgetting of space, which also greatly influenced the characteristic mathematical approach to time, as the algebraic time of calculations, by excluding the medium of the genesis of structure. This led to the severe reduction of the analysis of human cognition and, which is a greater distortion, of animal cognition, since humans can use logic and formal calculi as supplementary cognitive aids, which permit a biased grasp of at least a part of what is involved in understanding. Their analysis is just this part, which is least accessible to other living beings.

What marks an interesting historical reversal of this trend is that today we cannot study or seek to construct computers without taking account in a new way of considerations involving space and time. The geometric aspects of the structure of computers enters into the study of distributed, concurrent and asynchronous processing, which areas pose spatiotemporal problems of a kind completely foreign to the theory of Turing machines, the theory which dominated the study of computability from the 1930s to the 1980s. Computer networks and space distributed systems are the novel challenge. In these new areas the main problem concerns *time as related to the structural genesis and the constitution process*. This is a kind of time which involves space and which thus poses a new set of problems for computer science as well as for physics. Is this a further aspect of the new role of geometry in the study also of cognition? Should we think of the time of cognitive processing as an inherently distributed time?

Finally, where is the living system which does not exist other than in space and time? And this within a network of relations? Take the dynamical self-organization of ecosystems for example. Their genesis is above all a genesis of structure, from protein folding to the morphogenesis of an elephant; and their time is the history of a process of constitution. Dynamic irreversibility, gestalt, systemic unity and cohesiveness ... what happens to all these aspects of living systems which act in space and time?

3.2.2 *More on Biological time*

In the foregoing remarks, we have the outline of two ways in which we can regard *phenomenal* time as constituted, because it is jointly *construed* by us-and-the-world: it is, in turn, a constitutive element in our forms of knowledge of a reality-out-there, but one which must be endowed of structure to become intelligible. This time is at once a real and a rational time, remarkably, but not absolutely objective. It is the co-construction of the knowing subject and the world, as rooted in the regularities which we happen to see, we detect in the world – regularities which are out there but the explanation and the (scientific) objectivity of which are constituted in intersubjectivity – an intersubjectivity with a history. Let us now examine two forms of phenomenal time more closely, with the aim of suggesting a third.

The first form, which we called algebraic-formal, is that of clock mechanisms – the same clocks which the Enlightenment regarded as a possible model for the operations of understanding in general – and which later became the time of a (discrete state) Turing machine (see Chapter 5, for the "discrete *vs* continuous" issue in computational models of mind). A Turing machine tells time by the movement of scanning/reading its tape – to the left or right – tick-tock – like an absolute Newtonian clock. Nothing *happens* between one movement and another (to the left or right as the tape is scanned) nor can anything be said of their duration: these movements are the measure of time itself.

This notion recalls the time of myth inscribed in Homer. During the Trojan War, time is marked by the sorties of Achilles from his tent. Achilles leaves his tent, something (the War) happens; he re-enters his tent, everything stops, time stops. Achilles' motions provide the (only) scansion of time. Troy and the Trojan war (in the sense relevant here) lie outside the world – they exist in the realm of myth. Their internal time contributes to an extraordinary poetic effect. Turing and Homer are as one in this respect: the time characteristic of 1930s formalism is the time of algebraic-formal construction – the absolute time of a formal mechanism lying outside space, the time of "calculation-in-itself" is the time of one step after another in a void. In fact this time is secreted by the actions of a Turing machine viewed as a clock itself.

But Greek thought suggests a second representation of time as Kronos, son of Ouranus. Kronos (Time), derived from chaos and devouring his children, is "true," physical, time: the "paddle" of the real world. This version of physical time fits well with our analysis of dynamical systems displaying critical behavior (e.g., characterized by phase transitions). It is a time in a space – the space of the geometry of dynamical systems, a time recorded by their bifurcations, by their irreversible transitions from stable to chaotic regimes. Indeed it is the time of "the genesis of structure," of constitutive process, because a bifurcation, or a catastrophe, can depend on the entire history of a system, and not only its state description at a given instant.

To represent time as a linear continuum, the line of the real numbers, is very convenient; in many contexts one can choose no better model. But we here take up the reasons for the dissatisfaction with it which Hermann Weyl expressed in *Das Kontinuum* (1918a). Its "points" cannot be isolated in the manner of points on a spatial line because the present blends into, and indeed has no meaning except in conjunction with, the past and the future. While giving substantial contributions to the mathematical setting of relativity, Weyl recognized the limitations of relativity theory to represent time as an epiphenomenon, given that time is equipped with the same structure as the spatial continuum. Moreover, reversible time, due to the equations of relativistic physics, has nothing to do with phenomenal time as a mixture of experienced and rational time.

The time of dynamical systems theory and theories of critical states seems much better adapted to capture irreversibility than that of relativity theory. Moreover, the time of catastrophe theory can be given no meaning other than in space and in this respect it is like the time of relativity theory: firstly, bifurcations and chaotic behavior require space for their manifestation; secondly, there is no such thing as the time of a single isolated dynamical system displaying linearity in its bifurcations. No such system exists. The genesis of structures proceeds in parallel, through interaction of a plurality of structures (sub- and super-systems) in a spatial setting.

There are exceptions to the immersion of this second form of time, our genetic-structural time, in space. One could say, for instance, that the grammatical structure of natural languages, and other aspects of their structure possess a history and an existence in time without making reference to space. But language is an intrinsically intersubjective phenomenon – it is a plurality of speakers, situated and acting in space, which makes language possible. There is no language of an isolated speaker; language is always spoken in order to interact within a cultural ecosystem, which is often in friction with other cultures.

As the temporality of physical systems is associated with the genesis of structures in space through the interaction between systems, which are both dynamical and distributed, the *synchronization* of such systems becomes a central problem. Already in relativity theory, this problem shows up in the exchange of signals between differently accelerated systems. In computer science, this problem is partly bound up with the analysis of concurrency between processing units distributed in space. Both the time of Turing machines and that of Achilles' sorties *requiescant in pace*. Today we have a more "structured" time – that of a plurality of dynamical, distributed and concurrent (or more generally interacting) systems with their own local times, demanding synchronization. But if there is no time apart from this synchronicity, the same holds for asynchronicity, because it is already inherent in any "real" interaction between systems in a not purely local universe.

In summary, we have today a notion of time better adapted to our scientific understanding of the physical world: one enriched by the consideration of relativistic phenomena and (irreversible) dynamical systems. This time is essentially *relational* in character. Just as the absolute space of Newtonian physics no longer makes sense, so the absolute clock of the Turing machine, isolated in an empty universe, no longer seems to define an adequate representation of time. That would be akin to have the standard meter of Sevres isolated in an empty universe: in that universe there is no distance, just the meter.

There is yet a third form of time to be discussed and it is one appropriate for biology. The time in question is a phenomenal time, superposing experienced and rational aspects; it is constituted jointly by ourselves and the world, in the very acts of our intentional experiencing the world. That it manifests resistance to our *attempts* to grasp it is essential to its understanding. It is not to be thought of as "already there," yet it is not something arbitrary – because the regularities which supply us with clues and suggest how to speak about this time are certainly there; it is we, however, who choose to regard them and how.

In biology, matters are effectively more complex than in relation to physics, and one is obliged to move away from the idea that our brain (or any living organism) is a logical device or a programmable machine. First of all the "unity" and the "characteristic" time scales of living systems are related to the autonomy of the biological clocks of which we give a detailed account in the following section. This autonomy is even more striking than that of the mechanisms acting as clocks in the case of physics, because of the way in which a living organism strikes us as a unified individual. In physics, the present and future states of a system, and of the world as a set of dynamically interacting systems, depend "only" on past states. But the situation in the case of organic systems involves even more interactivity than that.

On the one hand, there are autonomous clocks appropriate to the individual system – its metabolic rate, its various biorhythms (heartbeat, respiration, etc). These are constants over ranges extending in some cases beyond entire species, even covering an entire phylum (the mammals for example). Evidently these clocks are far from being isolated systems – they regulate the functions of organisms in interaction with their environment; indeed their *raison d'etre* is to constrain and regulate that interaction.

On the other hand there is the phenomenal time of action within space on the part of this same living system – action characterized by aims and purposes, not least that of survival.

Before discussing this, however, let me review at least two further factors involved in the study of time in biology: the local time of each individual living being, its internal clock(s), which is re-established after any action affecting them (any action within the limits the organism can tolerate). Its clocks indeed exist precisely to permit and to regulate such interactions; a global time in which possible bifurcations in the system dynamics are determined, according to anticipatory capacities of the organism, by choices on which the possible future states (survival) of the system within its environment depend.

"Intentionality," as possibly conscious protension, is thus characteristic of biological time and it extends far below the threshold of consciousness, as is seen in the behavior of single-celled organisms which move in one direction or another to preserve their metabolic activity. This movement is one of the most elemental forms of choice: constituting bifurcations between possible directions (paths) of the system in phase space. In the case of human beings this choice is made on the basis of explicit awareness and conscious anticipation of the future. It thus depends on the range of possible future states considered.

It is thus a "contingent intentionality", related to contingent goals of the kind characteristic of different organisms. There is no organism or species without one implicit goal, that of surviving. But this finality is not metaphysical, rather it is immanent and contingent. If it were otherwise, neither the individual nor the species could long survive. It is essential to the preservation of living systems, from a single cell to multicellular organisms, as they are capable of future-oriented actions. Intentionality in the Husserlian sense of the term, involves an envisaging, a mental act consciously directed towards a target. Here it has a broader meaning: it is thus the *end result* of a network of interactions which plays a *constitutive* role in phylogenesis. Pachoud (1999) also suggests enlarging the Husserlian notion of intentionality in order to revitalize the phenomenological program.

Let us take an example from the study of primates. This example falls midway on the scale between the actions of an ameba in its metabolic responses and the conscious intentional behavior of a fully socialized human individual, or even the collective purposeful activity of an entire social group. This is the example: when we switch our attention from one point in our field of vision to another by a saccade (a rapid eye movement), the receptor field of the neurons in our parietal cortex is displaced suddenly, *before* the ocular saccade, in the direction in which we are thereby looking (Berthoz, 1997, p. 224). In other words the brain, in order to follow the trajectory of an object, or to escape the claws of a predator whose intentions it has "understood," displaces the receptor field of its neurons and anticipates the consequences of that displacement.

This is only one example amongst many which can be given of the role of anticipatory action of the future characteristic of living systems. We consider it of great interest because it is a form of intentional future-oriented behavior below the threshold of consciousness in animals, but very close to conscious movement. In fact, it seems that the glance actually produces a change in the biochemical (and hence the physical) state of the neurons, in the act of anticipating the future. This new state is imprinted on their structure – the new state in which they are then found does not depend only on their present and past states, but also on their anticipated future state.

In what follows, we will develop an analogy between local curvature in the Riemannian spaces of relativity theory and the locality of the internal time scales of living systems. A constant non-zero curvature provides for a local spatial scale linked to the (local) metric exactly as metabolic or cardiac rhythms appear to provide a time scale, more or less regular but local, observed in the individual system but common to the species or wider phylum. In contrast with the absolute locality of the metric of curved Riemannian space, local biological clocks are embedded in a wider ecosystem, and their *contingent finality* is not what it would be in the case of an isolated organism; rather they contribute to ensuring the relatively stable existence of the organism in a changing milieu. They *synchronize* it with similar systems and maintain it when in interaction with dissimilar ones. Whereas constant local curvature furnishes an invariant, local, metric element, independently of what goes on in the rest of the world, the internal clocks of living systems play a role in interaction. They aid in the establishment of a common time scale and they allow for the regulation and synchronization of other clocks within an ecosystem.

3.2.3 *Dynamics of the self-constitution of living systems*

Any individual organism, or any species, defines what may be termed a zone of "extended criticality," a critical state, as discussed above, yet extended in space and time (and in all pertinent parameters). We will present this notion in Chapter 6. As for now observe that it appears to be a situation impossible in physics, where "critical" states are generally unstable pointlike singularities. In this zone of extended criticality numerical invariants characterize the time scales of the autonomous system and reorganize the "unity" of the system in relation to heteronomy. When one examines a species embedded in an ecosystem much of the conceptual framework taken over from physics appears inadequate. Although the theory of dynamical systems has furnished some effective mathematical tools for biology, the study of a living system with the methods developed in mathematical physics has conceived the evolution of the system as taking place within a "fixed" field of force, or at any rate within a network of fields of force given at the outset. That is to say, the phase space does not change in the course of evolution.

A marble rolling in a cup is a simple classical system. Its field of forces: gravity, the geometric shape of the cup, frictional resistance – all already in place at the outset. The analysis of the ensuing oscillations follows very straightforwardly. In the case of more complex dynamical systems the mathematical analysis of their behavior may make reference to so many different forces that the majority of systems turn out to be intrinsically unpredictable. However, *qualitative* analysis allows us some remarkable insights into their possible evolution (the existence of singularities, bifurcations, attractors, and so forth), even in the absence of complete predictability. In the case of a living system a further factor is involved: the field of forces acting on the system is itself constituted in the course of the evolution of the system. In analyzing that evolution one may have to pass from one phase space to a completely different one.

Take a species within an ecosystem. Doubtless its interactions with the physical aspects of its ecosystem are determined by forces which relate to those aspects (e.g., gravitation, the physics and chemistry of the atmosphere or of seawater) but within an ecosystem one finds also other living beings. They react on the species in question. In fact species co-constitute themselves in conjunction with one another. And these other species were not necessarily present in the ecosystem before the one being studied, nor are they fixed and frozen entities. Their existence and evolution may itself depend on that of the species under consideration. Living systems in their interaction do not form a *given* field of physical forces – no minimum principle, no geodesic principle predetermines their evolution. For modern evolution (and we have for the present no better theory) they rather become more or less compatible with a situation, which living systems themselves will have co-constituted and co-modified, rather than with one given in advance.

Darwinian evolutionary theory refers to the combinatory explosion of life "in all possible directions." That is to say, no overall pattern of development in the system is predetermined, still less predictable, except in the case of small laboratory populations (e.g., of bacteria) under very controlled conditions. But, in general, evolutionary behavior is "just" compatible with (but could not exist without) the situation which it itself contributes to determining. Novelty arises on the basis of a given situation (which includes a genetic make-up) but also via the establishment of new patterns of interaction, the significance of which cannot be understood prior to their constitution. Gould mentions, for example, the tremendous role of "latent potentials" – illustrated for example by the double articulation of the jaws of certain reptiles 200 million years ago, which became the inner ear of birds and mammals. There was no *a priori* reason why things should have gone this way – no physical field of force and no genetic endowment on the part of reptiles imposed this development – it was made possible in the context of (indeed it was *co-constituted by*) an ecosystem. It would have been impossible to predict: it was not a necessity, like in the formation of a geodesic, but just a possibility to be formed, a generic path out of many compatible ones. The evolutionary explanation is *a posteriori*. We find ourselves further than ever from Laplace and there lies the scientific (mathematical) challenge.

Thus novel possibilities modify the field of forces set up by the living ecosystem. It is as if the cup in which the marble was set rolling assumed a shape (even a variety of shapes) from amongst all the physically possible ones, whilst the marble was in motion. But it is even more striking than that, for the marble too becomes extremely malleable whilst at the same time seeking to safeguard its unity and autonomy, just as all living individuals and species endeavor to do. Briefly, the biological "field" is co-constituted in time. In this respect it is something over and above physical fields; it *depends* on the latter of course, but is *not reducible* to them; at any rate we are a very long way from being able to produce such a reduction. The unification of biology with, rather than its reduction to, physics remains a principal aim. As we observed, physicists look for unification, not the reduction of relativistic and quantum fields. And, in order to unify, you need two theories. Moreover, it may be that this looked-for unification will come about from a quite different theoretical direction. It may be that an account of quantum phenomena will emerge within the framework of a general account of systems, including anticipatory capabilities. In this connection one will need to enrich the very concepts of "causal determination," "system," etc. Our aim at this juncture is a conceptual analysis, which pinpoints the parallels and divergences between new mathematical understandings, beginning with the issues of space and time. Clearly, the issue is theoretical, not "ontological": monist of the matter as we are, our main methodological assumption is that there is just matter, out there. However, what is a suitable theory for the living state of the matter? When moving from medium size physics to microphysics, we had to radically change the theoretical frame, and it is just a matter of "scale"!

What can be meant by a shift/enrichment of our concepts of "causal determination" and "system"?

Unfortunately, anyone who observes that the range of biological phenomena displays aspects which elude description in terms of current physical theory risks being branded an obscurantist and accused of believing in magic or to be a vitalist. The situation is not helped by the fact that one does indeed encounter terminology of a magical-poetical flavor in some writings on this subject. Confronted with this position, some tough-minded commentators cling to the notion of a deterministic program (in the sense of Laplace and Turing) and see it encoded in the DNA or into the brain, as the hardware on which the program, or rather a whole set of interlocking programs, is run. As for the brain, others take up the issue of quantum non-locality, locating its manifestation at the level of the microtubules of neurons and claiming that this will turn out to form the reductive basis of consciousness. Others again turn to the study of dynamical systems and take this as the framework for modeling the evolution of neural networks and the plasticity of their behaviors.

Clearly there are very important differences between these approaches. The first of them nowadays comes within a hairsbreadth of being a swindle. It has long been clear that we see less and less evidence in current physics of the kind of determinism embraced by Laplace and Turing, and even less in biology. This is not to deny the importance of both Laplace and Turing for rational mechanics and information sciences, respectively.

The second approach, concerning quantum entanglement in cells or in the brain, sets out a challenge to be taken up, but is currently lacking in experimental evidence, or in linkages between the scales of the structures and systems involved in the hypothesis: between the activity of neurons, which are very large scale structures, and that of the quanta, intermediary levels of description are altogether lacking. The third approach is founded on a strong body of evidence concerning the workings of the brain – the observed reinforcement of synaptic connections and, more generally, the effectiveness of the dynamical systems framework for the treatment of any interactive system. Here progress has been remarkable, yet the reduction is performed towards a specific physical theory: no novel conceptual unity is proposed. In particular, the formal neural brain has no body.

In these three approaches we also see the change of the notion of "determinacy." For Laplace (and for the sequential programming of computers) any deterministic system is completely predictable. [7] Within dynamic sys-

---

[7]Turing in 1935 himself demonstrated that a Turing machine was subject to a kind of unpredictability: one cannot decide the halting problem (whether the machine will halt or not in executing a given program). In fact we can decide no "interesting" property

tems theory, determinism does not necessarily imply predictability. Quantum physics introduces a further and deep-going modification of the concept, via its dual, the notion of intrinsic (non-epistemic) indetermination.

In less than 120 years, our notions of what it is that determines what, and our notion of what is a system evolving in time, have undergone a profound shift. But we still have no equivalent general notions in biology. We cannot say in what manner DNA determines the ontogenesis of living systems, nor in what way the state of a nervous system determines its later states. In an attempt to tackle these issues with the concepts of present mathematical physics, researchers have entered the conceptual kitchen, so to speak, and are busily drawing up a menu based on the recipes and cooking utensils they have already mastered, a menu drafted, where possible, in collaboration with the biologists. This is a fair, pragmatic attitude, but to make better progress we stand also in need of a robust notion of *biological field* or theoretical determination at the level of the organism, which is still lacking.

3.2.4 *Morphogenesis*

Let us now turn again to the the notion of space appropriate to the study of living systems. This analysis is one key aspect of the connecting tissue of this book as we try to comparatively develop it in the light of our understanding of the "construction principles" in mathematics and physics. Symmetries and geodetic principles will be discussed here and more extensively in Chapters 4 and 5, up to the extension of recent notions in physics (the physics of "critcality," Chapter 6), which singularizes the living state of matter.

---

of programs (Rice's Theorem: see Rogers, 1967). However, this unpredictability only becomes apparent in the limiting evolution of the system. The non-halting shows up only when a system performs infinitely many steps, and the undecidability of programs is a property of functions inasmuch as they admit infinite values and arguments. By contrast, the unpredictability of deterministic physical systems, investigated by Poincaré, is manifested already in finite levels. Given the initial state (defined with the due approximation), there exists a finite time (the Poincaré relaxation time) after which one *cannot* predict the state in which the system will be found. Despite the well-known undecidability results, classical computability theory thus conceives of a computation as a deterministic process of calculation and as completely constrained to follow a predictable evolution at any finite instant. Such a theory is a *logical* theory and Turing machines are *logical* machines. In neither case are they to be thought of as constrained by any physical limitations. The issues of determinism and indeterminism involved in the halting problem are issues of logic, not of physics and not subject to the hazards of physical approximation.

One of the areas in which we see the richest use of geometrical concepts in the study of living beings and their associated ecosystems is in the study of morphogenesis. In this we include the study of the evolution of the forms of living organisms and the influence of form on the structure of life in general.

"The stability of living forms is geometric in character" (Thom, 1972, p. 171). The topological complexity of a form is for Thom the locus of its "meaning" and of its organization. Thom assigns an almost exclusive explanatory role to topology: the topological evolution of the form of a living individual provides the explanation for its biochemistry, rather than the other way around (Thom, 1972, p. 175). The form in question contains information in two ways: it determines an equivalence class of topological forms under the action of a group of transformations; and it also supplies a measure of the computational complexity of a system via the number and evolution of its singularities. Here we can glimpse the idea of a "morphogenetic field" which fashions living systems, in the course of their phylogenesis as well as their ontogenesis. A *global* structure and operations of a *global* character occupy center stage in Thom's view. In the embryo, he emphasizes, we already have the global pattern of the organism, from which the specialization of organs and their function follows. As it has been said (Jean, 1994, p. 270): plants form cells, not cells plants.

But just what is this "morphogenetic field"? This expression could lead us astray if we think in terms of the physical fields. The morphogenetic field must be thought of as in some sense containing all the known physical fields at once, together with new fields characteristic of co-constituted organisms. In particular, each field – physical, biological, or cognitive, acts at a certain level of organization, *conceptually* independent of others: the phenomenal level and its conceptual structuring by our forms of scientific knowledge are completely distinct. However, the individual organism achieves *de facto* integration of this plurality of levels: its unity results from this integration of physical, biological, and cognitive levels. These different levels of structure and organization, analyzed by quite different scientific methods and concepts, interact with each other via spatial and temporal linkages. Each level displays plasticity with respect to the others.

However, no current physical theory supplies the concepts needed to describe these forms of mutual action, control and constraint, operating between the different levels: the ascending and descending linkages between them at all levels of the system and its biological, chemical, and physical components and subsystems. Cybernetics, the first theory of control automation, has certainly furnished remarkable models of the linkages involved in self-regulation. But these models have been located specifically at the physical level, and are constrained by the range of the theoretical tools they employ, whereas living systems establish linkages between conceptually wholly distinct levels of description (molecular, cellular, tissues, organs, organism). Typically, the mathematics of organ formation (phillotaxis, formation of lungs, vascular systems, ...) optimizes exchanges of energy. In contrast to this, the mathematics of cellular networks (neural in particular) deals with exchanges of "information" (mostly gradients of energy), a different observable. These two areas do not talk to each other, while it happens that cellular tissues are part of organs which are integrated in an organism.

Thom's analysis, subsequently enriched by the work of many other researchers, is also directed at the physical aspects of the topological plasticity of living forms, including those aspects induced by their "virtual" interactions. His work deals with an extremely informative *physis* of living systems, but still a *physis*. While, on the one side, it views the plasticity seen in the evolution of living forms as constrained by the dynamic fields operating in their morphogenesis, on the other hand, topological evolution is regarded as developing within a physical schema which takes no account of such phenomena as latent potentials or the combinatory explosion of life in all *compatible* directions. But, compatible with what? We do not mean compatible with the forces acting on a system at a given instant, but also with those it *will* experience. This poses a (mathematical) problem, which is at the heart of the evolutionary and developmental plasticity of living forms.

As it has been understood by those who have contributed to the most fully worked-out aspects of the theory of morphogenesis, namely phyllotaxis, it is possible to induce forms very similar to those seen in phyllotaxis by means of superconducting currents imposed on a magnetic field (see Jean, 1994, p. 264). For instance, the Fibonacci sequence, which is observed very frequently throughout the vegetable kingdom, can be reproduced by this method on any mesh of "soft objects" under repulsive forces and strong deformations (*ib.*, p. 265). In this sense, such an analysis does indeed consider living forms with respect to their being as purely physical systems – that is to say as bodies subject to the influence of physical fields. But although an important and necessary investigation, this is not exhaustive as an analysis of the forms of living systems.

Morphogenesis also has an important role to play in helping biology break out of the stranglehold of "genetic chauvinism" (that DNA provides a complete description of embryogenesis or even ontogenesis). The latter, in the writing of some authors, takes the form of a near maniacal expression of the Laplace-Turing vision of an absolutely deterministic causality, legislated in advance by the initial configuration of the system's components: some sort of homunculus residing in the genes, translated in the modern terms of a *coded* homunculus. This vision of a closed future is strongly rooted in currents relating genetics and sociobiology. This is unhappily congenial to religious believers in predestination: one just replaces God for evolution as the typist on the molecular keyboard of the genetic program and ontogeny is completely understood (including marital fidelity, see Young *et al.* (1999)).

In contrast to such a picture, Atlan replied: "the program of a living organism is everywhere except in its genes": certainly the patterns and the forms seen in phyllotaxis are not entirely in the genes. They are also in the structure of space and time and of physical matter and energy. The genes do *not* contain all the information on the symmetries which are set up in a system in interaction with its environment, such as are observed in crystals and minerals (Jean, 1994, p. 266). The so-called "program" for the development of an organism is to be found in the interface between its phylogenetic record (its genetic legacy) and its physical and biological environment (its ecosystem).

An example of the greatest importance is provided by the brain, which in the course of ontogenesis manifests a developmental pattern, which is both Darwinian and Lamarckian. The immense number of possible connections among its neurons (each one of around 100 billion neurons has up to 10,000 synaptic connections, maybe more) could not be (or at any rate very little of it could be) encoded in the genes. Of the numerous connections established very rapidly during the growth of the fetus or the newborn child, most disappear through selection effects (Edelman, 1987). On the other hand, throughout the course of our entire life, stimuli lead the brain to establish new connections and reinforce existing ones, jettisoning and replacing existing connections as it does so, selecting certain neurons and leaving others to die off. Cerebral plasticity, at all levels, is at the heart of the continuity between phylogenesis and ontogenesis, and is what permits continuing individual identity: "the structure of the nervous system carries the material traces of its individual history" (Prochiantz, 1997).

3.2.5 Information and geometric structure

In an epoch of free-floating bits, the picture of information (even of intelligence) as purely a sequence of bits enjoys great currency. The digital encoding of information is of great effectiveness for certain purposes: once encoded, such information can be safeguarded and transmitted with unrivaled accuracy and speed. No method is superior to that of bit-storage in the construction of digital computers and the networks they form, which are now in the course of transforming our world.

Moreover, a number of notable mathematical results of the 1930s demonstrated that *all discrete encodings and their effective treatment are equivalent*. Kleene, Turing, Church *et al.* demonstrated the equivalence of (very) different formalizations of "computability": the numerical functions calculated by using the systems of Herbrand, Curry, Gödel, Church, Kleene, and Turing were the same. By means of an astonishing philosophical sleight of hand, trading on the surprising and technically difficult nature of these results, and influenced by the surrounding intellectual climate of formalist and positivist ideas, the claim was *later* made that *any physical form in which information is processed,* and thus any biological form of information processing or any form of intelligence, *can be encoded in any such formal system,* thus it can be encoded in the form of the strings of 1s and 0s used in the memory stores of digital computers, see Longo (2007) for more on some parodies of the Church-Turing thesis.

A quite different way in which information can be thought of as structured, one involving geometric principles, is through equivalence classes of continuous deformations. These provide for the transfer and processing of much of the information essential to the make-up of living beings and, more generally, of physical systems. Continuous, differentiable or isometric transformations and the regularities they preserve or fail to preserve may help to structure and make intelligible living phenomena, as can be seen not least in the geometric structure of DNA or of proteins and their evolution. To these transformations the discrete and quantitative structure of bits of information serves as an addition; bits behave as singularities and thus as a possible measure of the topological complexity of the geometric structure. Information has both a qualitative and quantitative nature. The concentration on only its quantitative, digital, nature has become a severe limitation when information is assigned an explanatory role.

A frequent reaction is: yes, granted the role of these kinds of transformation and this kind of continuity, nonetheless, in the last instance, the geometric structures involved are reducible to very minimal discrete units. Such a reaction hides many problems. Physical and mathematical principles prevent our modeling continuous and three-dimensional information in discrete form. The notion of (Cartesian) dimension does not apply in the discrete frames of codings and programs. Infinite discrete structures, the natural numbers $\mathbb{N}$ in particular, are isomorphic to any finite product: $\mathbb{N} \equiv \mathbb{N}^m$, by a computable isomorphism. That is, any finite string of integers can be encoded as an integer, and this is crucial for Gödel's and Turing's approach to computability and their applications.

These isomorphisms make no sense in mathematical physics, as they would simply destroy most of its theories: dimensional analysis, that is the analysis of the number and type of variables in a function $f(x_1, \ldots, x_n)$, is crucial in physics and theories radically change when changing dimension. From the analysis of heat propagation (Poisson equations), whose characteristics are very different in one, two, or three dimensions, to the "mean values theories," which differ radically from two, three, or four dimension, and a lot more. Not to quote relativity theory where the unified *four*-dimensional structure of space-time is crucial or string theory, where intelligibility is given by moving to 10 or more dimensions. And mathematics proves it beautifully, in relation to the "natural topology" on $\mathbb{R}$, the real numbers, that is in relation to the so-called Euclidean or interval topology. Recall first that the interval topology is "natural" since it comes from physical measure, which is, by principle, an interval. Then, and this is fantastic, one can prove the following:

if $A \subset \mathbb{R}^n$ and $B \subset \mathbb{R}^m$ are open sets and $A \equiv B$, then $n = m$.

This theorem says that dimension is a topological invariant, when one takes the natural topology in a space manifold, in the sense of Riemann. This result is false when considering, for example, the discrete topology on $\mathbb{R}$, or, say, a weakly separated topology. This is a remarkable connection between mathematics and physics, via measure: the interval of physical measure, which yields the naturality of the Euclidean topology on the reals, makes dimension an invariant and gives a mathematical meaning to its central role in physics. Thus, there is a mathematical gap between the continuum and the discrete, which is not, in general an approximation relation; instead, they provide different mathematical intelligibility of phenomena.

Now, the smallest living phenomena comprise dynamical systems (thermodynamic systems, systems with critical points). And here is a further reason for a discrete mesh laid down *a priori* to be sufficient for their analysis. This is because sensitivity to initial conditions typically generates far-reaching consequences at or above the threshold of discernibility, triggered by a variation below that of the smallest interval of the measure.

But what kind of discreteness are we really talking about here? On the assumption we can push the encoding right down into the microphysical realm (so as not to have to cover the whole earth with processors) it seems the discreteness in question will come from quantum physics and will arise at the scale of the Planck length. One then encounters a fallacy, well explained in Bitbol (2000a/b), of the same stamp as that involved in the case of the formalist and mechanist reduction of mathematics to formal manipulation (processing) of discrete symbolic inputs. The reference to well-defined and ultimate discrete level of "material points" is the *sine qua non* of Laplacian mechanics. No such appeal is possible in quantum theory, because it is a theory of continuous (quantum) fields, where Planck's constant, the only possible referent for these notions of ultimate discretization, has the dimensions of an action (energy times time), which may be understood as a "fibration" orthogonal to the continuum of space-time. Moreover, quantum indetermination and the epistemological debate which has raged around it involved assigning a role to the knowing subject of a profoundly *anti-mechanist* kind. In approaching this topic armed with a digitized version of microphysical structure one is lending support to the myth of Laplacian mechanism, as a matter of the bit-strings programmed by formal laws of thought. This is the opposite of the the paradigms proposed by quantum mechanics.

We cannot renounce the mathematics of continua. The deformation of geometric structure can reproduce information in analogue form. And *the analogy involved is intentional* – one *chooses* what to represent or reproduce in analogous form, one selects those aspects of the original form, which are to undergo processing or simulation. The choice of analogy is the outcome of a controlling vision or an aim, conceptually appropriate to living systems, of a kind which is missing in physics, where the phenomenal arrow of time is oriented without backwards linkages; see, nonetheless, Novello (2001). The fact that the reproduction and transformation of information via geometrical forms is analogue in character and may better accommodate intentionality, appears to be just what is crucial for biological representation. The eventual greater instability over time of geometrical forms by comparison with binary bit-strings is actually an *enriching* factor, because it corresponds to the possibilities of evolutionary change. The analysis of the geometric structuring of living systems (particularly of the brain) permits us to grasp a factor essential in information: selective analogical simulation carries with it evolutionary possibilities. By contrast with this, the perfect stability of bit-by-bit information processing renders such an elaboration impossible (whether this is a practical impossibility or one of principle is unclear).

To conclude: in studying living phenomena, from the most elementary systems all the way up to cognitive agents, it is not so much a matter of denying the important role played by formal and mechanical aspects ("bits" are key singularities in relation to information) but rather of enriching this analysis through the phenomenal richness of geometric structures and their effectiveness in information processing. Once again, formalism and mechanistic physicalism are seen to be not a variety of scientific reduction of the kind we should expect to meet in scientific practice, but rather a philosophical monomania, which has lost touch with the plurality of forms accorded to our knowledge of and interaction with the world.

We mentioned the areas of morphogenesis and cellular networks as very lively but far apart research areas. Work on both morphogenesis and the architecture of dynamical neural networks (see e.g. Hertz *et al.*, 1991; Amari and Nagaoka, 2000, for the latter), despite its incompleteness, arising from the limitations of a physicist's, though neither formalist nor mechanist, standpoint, at the very least suggests the richness of the geometrical (continuous) structures implicated in any account of the organization of living systems.[8]

---

[8]The constitution of geometrical patterns of neural networks is a typical result of this complex compositional activity and permanent dynamism, but it is not the only result. See for instance the remarks in Edelman (1987) on the fine structure of synaptic connections and other aspects of neural plasticity, which go beyond those modeled by the dynamics of neural networks. Consider also the interactive genesis of the forms of such networks in the context of an ecosystem. Here the stimuli are of a physical or biological nature, grounding the mental activity in the material of living systems. Animal intelligence involves a dynamic of such forms distributed over many different levels of structure (from that of proteins to that of synapses to that of entire neural networks) all of which are in mutual interaction. Its unity is that of a subtle and complex kind of interacting and self-reacting field which is still very difficult to grasp.

3.3 Spatiotemporal Determination and Biology

3.3.1 *Biological aspects*

The question of space has played a very important, even a foundational role in biology: one which has not always received due appraisal. Take for example the concept of *milieu interieur* (internal environment) introduced by Claude Bernard, which allowed an essential topological separation between the interior and exterior of an organism. Consider also the question of chirality in biology, highlighted by Pasteur. In the wake of his experiments on the tartrates and the manner in which their biological activity differed depending on whether they coiled to the left or the right, he stated unhesitatingly: "Life, as it is manifested to us, is a function of the asymmetry of the universe and a consequence of it."

Indeed Pasteur anticipated both developments in his own field of scientific inquiry and, *mutatis mutandis*, what later came about in physics with the discovery of the asymmetry of matter and anti-matter, which cosmology now views as the precondition or the existence of the universe and of the actual material structures we see all around us.

Biological structures are subject to organizing processes leading to the emergence of complex forms, such as those studied in developmental biology; furthermore, they display physiological functions which sustain the part/whole mutual dependence, which mediate their integration as organisms and regulate the linkages between the different levels of organization typical of organic existence. These facts clearly have a connection with theories of the critical behavior of dynamical systems, such as that seen in phase transitions.

It was not by chance that the first mathematical models of biological systems appealed to and borrowed from those of thermodynamics, in particular models of cascade effects in bifurcations of thermodynamical systems (see Nicolis, 1986; Nicolis and Prigogine, 1977), followed by models of emergence of self-organized critical behavior (see Haken, 1978; Kauffman, 1993; Varela, 1989), and application of fractal geometry (see Mandelbrot, 1982; Bailly *et al.* 1989; Bouligand, 1989) and chaotic regimes (Babloyanz and Destexhe, 1993; Auger *et al.* 1989; Demongeot *et al.* 1989) to an organic context. Alongside these developments, it had been clear that the character and genesis of processes of formation could in many cases be modeled using the elementary theory of catastrophes (Thom, 1980), and, more generally, singularity theory. What clearly shows up in the analysis of self-regulation and homeostasis (but also in the analysis of pathology and death) is what may be termed the "extended criticality," i.e.the enduring sensitivity to critical parameters of systems in that situation – a situation which is limited in spatial and temporal extent, but which nonetheless is extended (see Chapter 6).

As we shall see, this notion of extended criticality makes no sense in current physical theories where critical transitions are always mathematically considered as pointlike phenomena. Yet, it may help us understand the living state of matter by extending physical concepts, sometimes by dualities with respect to physics, as we will argue. For example, recall that in the framework of quantum theory, energy and time are conjugate variables. But an asymmetry nevertheless holds between them. While energy is a well-defined observable of the quantum system, associated with a Hamiltonian operator, time appears only as a parameter: one seemingly less essential and less well incorporated into the theory. In biology we seem to have the inverse: it is the time characteristic of biological systems (an iterative time which regulates biological clocks and internal rhythms), which seems to be the essential observable; whereas energy (the size or weight of an organism for example) appears simply as a parameter, in a sense an accidental parameter at that. In this sense one might even say that biology, relative to the energy/time conjugacy is quasi-dual to quantum mechanics. And perhaps this duality, if further developed, may yield some more understanding about life (see Bailly and Longo, 2009 for some applications of this idea).

3.3.2 Space: Laws of scaling and of critical behavior. The geometry of biological functions

Since the pioneering work of D'Arcy Thompson (1961), recent studies (Peters, 1983; Schmidt-Nielsen, 1984; West et al., 1997) have shown that numerous macroscopic biological characteristics, as distinct from genetic traits at the biomolecular level, are expressed at the same scale across the range of entire species, indeed across genera, taxa, and in some cases the entire animal kingdom. This scale-invariant parameter picks out the organism by its mass W or in some cases by its volume V. Furthermore, characteristic time scales for organisms (lifetimes, gestation periods, heart rates, and respiration) all seem to obey a "scaling law."[9] They are typically in

---

[9]These laws concern the structural and functional consequences of the size (weight) or scale-changes in organisms.

an exponential dependence of one fourth of the mass (time, T, goes like about $W^{\frac{1}{4}}$). Just as these frequencies scale as $W^{1/4}$, metabolic rates typically scale in a ratio of $W^{3/4}$ and many other properties display similar scaling. Such scaling laws call to mind the behavior of dynamical systems where a critical transition to a regime is associated with fractional exponents of some key parameters. What differentiates one group of organisms from another is simply the value of the ratios seen in the expression of these scaling relations. These remain the same across numerous species and even across much wider biological groupings. Perhaps the most spectacular example is that of lifespan, which is in the same ratio to body mass.

Other kinds of scaling laws – allometries – link geometric properties of organs (as distinct from organisms) across numerous species, or within a single organism at different stages of its development. Here, however, our principal point is bound up with the display of fractal geometry in certain organs, engaged directly in the maintenance of physiological functions, such as respiration, circulation, and digestion. This fractal geometry appears to be the objective trace of a change in the level of "organization" and of top-down regulation of the parts by the whole, in conjunction with the bottom-up integration of the parts within the whole.

The fractal geometry in question falls into two distinct kinds. On the one hand examples are seen in the interfacing membranes of the organism. The metric dimensions of these are between 2 and 3. Examples are the membranes of the lungs, the brain, and the intestines. The other class is formed by branching networks, such as the bronchial tubes, or the vascular and nervous systems. Here the metric dimensions of the extremities may be greater than 2. These fractal geometries permit the reconciliation of opposed constraints associated with spatial properties. On the one hand, because the organs involved are engaged in the regulation of exchanges such as respiratory or cardiac function, their effectiveness and their corresponding size must be maximized in order to support and to fine-tune these exchanges; on the other hand the fact they are incorporated into an organism containing many parts means that their bulk must be minimized to ensure their overall viability. To the extent that organs are clearly individuated and allotted wholly to certain specialized functions, they must present a certain homogeneity throughout their spatial extent. These various constraints are clearly antagonistic and only the fractal character of the geometry of the organs in question allows them to be reconciled.

Another aspect which raises interesting questions is the intrinsic three-dimensionality of living systems. If one inquires into the abstract possibility of developing biology in dimensions other than three, one recognizes that the choice of three dimensions again allows the reconciliation of antagonistic constraints. On the one hand it is required that the organism has sufficient local differentiation to permit different concurrent functions across its whole structure; on the other hand it needs to be the site of sufficient internal connectivity to co-ordinate the activity of all its parts. In a space of only two dimensions, if the differentiation were sufficient, the connectivity and co-ordination between the different parts could not be established because the required connections of the components would intersect so greatly as to disrupt their separated functioning. On the other hand, in a space of four dimensions, the degree of possible connectivity is clearly greatly enhanced, but it is known that in four dimensions "mean field theories"[10] become applicable and the constraints of local differentiation become insufficient to allow for the establishment of systems stratified into different levels of organization. As a consequence, the value of observables on one point is given (almost everywhere) by the mean value in a neighborhood. Thus, barriers, membranes, singularities, ... become more difficult or impossible (we will go back to this in Chapter 5). Development of the system in three dimensions serves to reconcile these constraints at the cost of producing fractal geometries (their emergence reflects the existence of certain dynamical attractors). Note that these considerations concern the internal space of biological systems – they have no bearing on the dimensionality or topology of external space.

Following on from the earlier presentation of the notion of space in physics in terms of fibrations (carrying the internal symmetries of the system) over a base space (the external space-time) one might consider the existence, in similar terms, of internal "spaces" associated with different levels of biological organization. These should be distinguished from the different levels of scale structure in physical systems. What is distinctive in the biological context is the way these levels are connected with the regulation of the lower by the higher level of organization and with the manner

---

[10] Intuitively, mean field theories (for example Landau's theory of ferromagnetism) are theories in which an approximation consists of replacing the effect of an element of the system, in the ferromagnetic case a spin, by the sum of all the individual interactions due to all the other elements by a "mean field" integrating their effects. These theories become better adapted for the description of the system in question the higher the number of near neighbors of any given spin is raised, because then the spin in question is better seen as the mean effect of these others. In a four-dimensional space their number is sufficient to provide a model of a mean field theory.

3.3.3 Three types of time

To a first approximation, in describing the actual state of an organism (or a population) one can consider two types of biological temporality, jointly implicated in its survival. The first type, which carries echoes of time as seen in classical physics, is associated with the stimulus-answer coupling between an organism and its environment. It is manifested chiefly by relaxation processes (in the quasi-canonical form $e^{\frac{-t}{\tau}}$ and exponential combinations thereof).[11] The second (of a very different nature) is associated with internal clocks which administer the biorhythms of a living system and ensure its continued functioning (Glass and Mackey,1988; Reinberg, 1989). It takes the form e^{iwt} and its combinations.

But the most important aspect of biological time is perhaps irreducible to these distinct forms: the internal "temporality" of organisms is iterative rather than historical. The measure of duration in this internal time is no longer a dimensional magnitude, as in physics, but rather a pure number registering the iterations already effected and those still remaining for an organism which experiences a finite number of these within a range fixed in advance, depending on its class. Thus, all the mammals, from mice to elephants or whales, form one such class. This is characterized by the number of heartbeats per average lifetime (around 10^9 for mammals) or the number of corresponding breaths (around 2.5×10^8). The variation in these frequencies between species is traceable to a single parameter – the body mass of the average adult. This striking trait is directly connected to the scaling law mentioned earlier.

The importance of this aspect of biological time is emphasized by recent attempts to re-think the principal features of evolutionary theory in terms of the "living clocks" approach (see Chaline, 1999). By interpreting evolutionary transformations in terms of their synchronic and diachronic effects, this approach concerns both the developmental level of individual organisms and the evolution of species.

---

[11] The simplest example of a relaxation process is the return to equilibrium of a system that has been subjected to a small perturbation. The speed of return is proportional to the departure from equilibrium the system has undergone. If P is a quantity of equilibrium-value p, with $P > p$, then $dP/dt = -r(P - p)$, where r is the inverse of a time; this leads to an exponential decrease in the departure of the system from equilibrium with time. The inverse of r is the characteristic "relaxation time" of the system.

But as we have proposed in Section 3.2, it seems that in biology it is necessary to take into account a *third* type of temporality, connected with what we may term "contingent finality." This expression means a degree of (non-reflexive) intentionality which may help to explain the evolutionary and adaptive aspects of living systems. It is an *anticipatory* form of temporality, linking the current state of the organism and the future state of its environment, to which the organism contributes by its behavior; and it is a form *specific to biology*, termed as "teleonomic" by Monod, which arises in connection with a coupling between the rhythms registered by inner biological clocks of the system and the stimuli-responses the system undergoes while interacting with its environment. (Of course, this coupling may introduce delay effects.)

Unlike the two previous dimensions, this "third dimension" corresponds to aspects of biological systems not seen in physics, in that it concerns the variety of integration-oriented factors involved in determining the present state of the system with those involved in determining its future state.

Perhaps this is another reason why, as we have already underlined, one cannot define the notion of a trajectory traced out in the course of the evolution of a biological system in the manner one does for phase spaces in physics. There is no scope in the biological case for the application of a geodesic principle which extracts and determines one trajectory, that obeying an extremal principle, from amongst all the virtual possibilities. It seems the logic of biological systems operates in a quite different fashion, in a manner designed to display a *Bauplan* selected by external criteria.

Gould claimed in his account of the organisms found in the Burgess Shale that it seems as if all the virtual possibilities, or every possible pattern of development, saw the light of day. Here it seems we are dealing with criteria operating not so as to secure the emergence of a single possible form of the system (as with minimum principles in physics) but rather so as to secure the elimination of impossible forms so as to produce a maximal variety of system structure within a limiting "envelope of possibility."

We cited the fact that biological systems, rather than following an evolutionary trajectory (tracing out a geodesic) explore all the possibilities compatible with their continued existence in a manner at once passive (i.e., subject to the effects of natural selection) and active (in modifying the environmental conditions in which selection operates). One can find analogies in physics for the first aspect, but not at present for the second, which appears specific to biology. In quantum field theory, path integrals (Feynman integrals) are constitutive of entities, still not yet everywhere well-defined, seeking to take account of *all* the paths (with their appropriate weighting) from the initial to the final state and not only privileged trajectories such as geodesics, albeit the probability of non-geodesic paths is very low. It is rather as if we could take account, in the case of biological systems, of all the transformations, which a given form of the system could undergo (together with their probabilities), as suggested by Gould's description of the Precambrian explosion to which the fauna of the Burgess Shale bear witness. All the quantum paths which determine the state of a system in the Feynman formalism are located in spaces (and involve modes of interaction) completely defined in advance and which do not really depend on a specific path (even if they can be regarded as depending on the final state, once reached). In biological systems, by contrast, any stage in the evolution of a system (even of an individual) modifies the conditions in which all subsequent stages are produced and these conditions are not defined in advance.

Without seeking to formulate premature conclusions, we can nevertheless draw some lessons for our understanding of our notions of space and time, in the light of recent developments in theoretical biology. The distinction between internal and external space is connected with the distinction between the autonomy of an organism (the homeostatic stabilization of its functioning and its identity) and its heteronomy (its dependence on and adaptation to its environment).

Equally, the internal/external articulation of a space physically determined in structure and another space, determined by the functionalities of the organism and its complex morphology, leads directly to issues of the relationship between the genetic inheritance of an organism and the epigenetic factors involving its interaction with its environment in the course of its typical development. The essential new element distinguishing the situation in biology from that in physics lies in the fact that in biology the articulation of this internal/external distinction applies also to time and in a crucial way, involving the relationship between the different types of temporality: one of them akin to that in physics and with the character of a dimension, the other specifically biological, iterative and expressed in pure numbers, which seems to play a quasi-constitutive role with regard to our concept of a biological system, in that it supplies the basis for the characterization of the invariants operating in the definition of equivalence classes in the biological setting (the invariants of mammalian biorhythms illustrate this well).

The manner in which the *concepts* of space and time are treated is, once more, fundamental to the constitution of biology as a science. Beyond the analysis of perception, any epistemological account of the status of such concepts – to the extent it is based on the results of natural science and aims to be objective – cannot ignore their role in the framework of theoretical biology.

3.3.4 *Epistemological and mathematical aspects*

It may be instructive to run over the epistemological ramifications of space and time by means of an analysis "transversal" of the theoretical frames of reference considered so far. Three pairs of concepts appear important for such an analysis. First, *local vs global* concepts in relation to space; second, *iterative vs processual* aspects in the understanding of time (with an aside examining how both these pairs of concepts are intimately bound up with the topic of *causality*). Third, *regular vs singular* in connection with our system of representation and reference.

In fact, by considering physical and biological aspects of space and time, one cannot evade the epistemological issues involved in their purely formal definition. Spatial and temporal concepts are "abstract" concepts in two different senses of that term. The first relates to the process of abstraction which takes its cue from common features of the theoretical treatment of space and time and the second, concomitant, sense relates to their being formal, quasi-*a priori* notions which come to be imposed as the result of that process of abstraction, as an intrinsic component of our notion of objectivity itself.

(I) LOCAL *vs* GLOBAL. In passing from relativistic to quantum theories and then to general dynamical systems theory, and finally to biology, we recognize a shift in the relative importance and pertinence of the local/global opposition. Broadly speaking, it is a shift from the local to the global. Despite the stress on a global interactive point of view seen in Mach's principle,[12] general relativity completely preserves the principle of locality inasmuch as it is essentially and exhaustively expressed through partial differential equations, where variables are meant to depend on "points" of the intended space. In this respect, it lends itself to an interpretation in terms of *local causes* propagating, typically, within the light cone.

---

[12]The local motion of a rotating reference frame is determined by the large scale distribution of matter, or, in Einstein's formulation: the overall distribution of matter determines the metric tensor, thus the local curvature of space.

Quantum physics can equally well be presented as a theory of local interactions and their propagation by state vectors. The Schrödinger equation and that of Dirac are just as much partial differential equations as those of Einstein. Hence, they could be taken to involve a notion of causality of the same apparent kind. But quantum measurements on the one hand, and non-separability on the other, stand in the way of a completely local interpretation of the theory. Classical causality is affected by the fact that measurement leads to intrinsically probabilistic results while non-separability disrupts any purely local representation of the propagation of effects.[13]

The case of theories of the critical states and dynamical systems takes us a step further: here non-locality plays a twofold role. Firstly the fact that interactions can now take place at long distance leads to correlations becoming infinite. Local variations and effects lose their relevance both for analysis and measurement in favor of the global behavior of the system. This even reaches the point that our notion of what counts as an object needs to be re-defined. Furthermore, this global behavior is itself governed by critical exponents and scaling laws which are in no sense local, since they are dependent on the dimensionality of the embedding space and on an order parameter.

A concomitant of this situation is that the usual notion of causality, even when there exists a linear correlation between cause and effect – small causes giving rise to small effects, is undermined. In critically sensitive systems, infinite effects can arise from finite causes which lead to discontinuities in the evolution. Curie's principle (that symmetry of causes is mirrored in symmetry of effects) is thus called into question, at least for systems displaying singularities and discontinuities in their behavior, by the symmetry breaking which accompanies phase transitions.

In biology, locality seems pertinent chiefly to the description of underlying physico-chemical processes, while the definition of biological systems and their manner of functioning involves global concepts associated with the fundamental non-separability of living systems and their complexity. This global level of structural organization becomes decisive for the representation of processes of regulation and integration which stabilize the functioning of a biological system. To this there corresponds a more complex notion of causality involving an entangled and interactive hierarchy and its associated "agonistic or antagonistic" effects. In brief, the notion

---

[13]Theories of hidden variables, proposed to overcome certain of these "a-causal" aspects, are themselves non-local, see Chapter 7.

of local causality cannot be called into question without the global notion being affected as well: in an organisms almost everything is correlated with almost everything, and this by its global unity. From the point of view of organs, then, the global notion is associated with the "contingent finality" of their activity: their local activity is understood as a "purpose" from the point of view of the organism. This also opens the way to a distinction – one meaningless in physics – between the normal and the pathological. A locally pathological mode of functioning can co-exist with the preservation and global functioning of an organism.

(II) ITERATIVE *vs* PROCESSUAL. Relativistic theories, with their characteristic metric structure, display an almost completely spatialized type of temporality. They introduce the concept of an event as a "marker flag" in a generalized space. Only physical causality – the fact that interactions between point-events cannot propagate outside the light cone, or to reformulate that requirement from a mathematical standpoint, the fact that the signature of the metric is fixed – serves to introduce a distinction between spatial and temporal dimensions. Conceptually, this distinction is intimately bound up with the fact that from the viewpoint of the symmetries of the system, Noether's theorem classifies time as a conjugate variable with respect to energy (or conversely, views energy as the conjugate of time), just as the spatial variables are conjugate to the components of the momentum. But essentially, relativity, via its group of general transformations, treats time as a notion of the same kind as space.

Contrasting with this situation, in quantum theory time is treated as a simple parameter. Its status as the conjugate of energy is preserved, as in the Heisenberg indetermination relations, but it does not appear as an observable of the theory. Moreover, it seems that certain phenomena – those connected with quantum state transitions or even measurement – do not easily lend themselves to an interpretation in terms of temporal concepts of the kind connected with our experience of passage and duration. One sees something similar in the apparently instantaneous connections associated with the behavior of non-separable quantum systems. This further stresses how greatly the conceptualization of causality is bound up with that of time.

In theories of dynamical systems, time regains more classical characteristics, but ones made apparent in connection with different aspects of phenomena than in the classical case. Besides being the time of events (as state transitions), it plays further roles, distinct from the role it plays as a parameter: in the definition of stability, in the definition of irreversibility, in the characterization of attractors (asymptotic behavior, as seen in fractal geometries, etc.), time assumes a constitutive character, that is these very notions make sense only with respect to time evolution.

As we have hinted, in biology, temporality displays two quite distinct aspects: the external time, i.e., the relaxation time of stimulus and response, of functional adaptation to an exterior environment, and the iterative time of pure numbers associated with internal biorhythms involved in the regulation of physiological functions. The corresponding notion of causality, adaptive and intentional in character, seems closely connected with the mutual articulation of these two aspects.

(III) REGULAR *vs* SINGULAR. Relativistic space-time is "regular," continuous and differentiable, and singularities (whether of the Schwarzschild or the initial, Big Bang singularity) play a quasi-incidental role, which assumes central importance only in certain astrophysical and cosmological contexts.

The situation is different in quantum physics, where the regularity of certain spaces is associated with the discretization of others and where space-time structures can be envisaged as fractal at very small scales. The regular/singular couple carries the traces of the old debate about the interpretations of the theory, namely in terms of fields or in terms of particles, as singularities.

In theories of "critical" systems, the interest in singularities is accentuated (see Chapter 6). They are associated with an increase in complexity, and also with the particular consequences of non-linear dynamics. In fact, these theories are essentially singular since critical situations all involve singularities (divergence, discontinuity, bifurcations). The mathematics of singularities (singular measures, catastrophes) plays a predominant role in modeling the behavior, which gives rise to complexity. Nevertheless, it appears that the outcome of these features is in fact a new form of regularity, one located at a more general level of analysis, revealed in laws of scaling and leading to a universal classification embracing very different systems, which nonetheless manifest identical behavior with respect to the singularities in their dynamical evolution. The critical transitions in these systems are typically restricted to a very narrow range, even to a single point in phase space, a single value of the control parameter, and on either side of this very narrow critical zone regular behavior of the system once again becomes dominant.

Precisely this last aspect seems to contrast with the position prevailing in biology. Organisms and ecosystems can survive and maintain themselves within a range of values of critical parameters – an extended zone of criticality. Exit from this zone implies the death of the organism: its underlying physico-chemical structure can no longer sustain biological functions. Any biological system behaves in a manner characterized by a *dense* distribution of critical points in the space of control parameters, and not a discrete or isolated one. Homeostasis then, corresponds to a sort of structural stability of the trajectories, relative to the attractor-basins of the dynamics.

3.3.5 *Some philosophy, to conclude*

Space and time, especially as they feature in the framework of modern physics, are neither *objects* nor *categories*. To recall Kant's formulation (Kant, 1986a) they are "*a priori* forms of sensible intuition," and as such the preconditions of any possible experience. In the light of the most profound analysis our current physical theories allow us to make of them, they seem to reflect the mathematical structure of a group and a semi-group, respectively.[14]

Indeed, the mathematical properties postulated for space, inasmuch as it is the medium and support of displacements in general, necessarily connect with and exemplify the group structure. Given the tight connection between the group structure and its associated equivalence relations, an abstract, epistemically basic frame of reference emerges which acts as a kind of pole of attraction for any representation of objectivity, namely the frame: <space, group structure, equivalence relation>.

Similarly in respect to time, the property of possessing an orientation – "time's arrow" as the index of change – is reflected in the abstract structure of a semi-group. That structure can be put in correspondence with an order relation. This leads us to the recognition of a second epistemically basic frame of reference: <time, semi-group, order relation>.

To repeat, for the avoidance of all confusion: the space and time which feature in these frames are not so much an aspect of entities with an intrinsic nature, but rather of the conceptual grid presupposed by any natural science: they are conditions of possibility rather than of concrete actuality. If this approach is correct, these two "poles" leave their mark in our notions of permanence and change, stability and evolution, identity and

---

[14] A semi-group is a set endowed with an internal operation which is associative; if the set contains an identity with respect to the operation and each element possesses an inverse, then it is a group. By this latter property, groups may be associated with symmetries, see Chapter 4.

differentiation. They delimit the "field" of preconditions for any natural science, to the extent that the phenomena studied therein are manifested in a spatiotemporal setting.

This *Tableau General* is still incomplete, notably in the life sciences. But what clearly stands out is that contemporary theories across a whole range of scientific domains involve conceptions of space and time that have not yet been fully stabilized or clarified (although in physics supersymmetric string theories are seen by many as leading to unification). Will we continue to distinguish space radically from time despite the merging of their status in the setting of relativity theory? Shall we continue to refer to a single notion of space in view of the variety of topological and other structural properties (compactification, non-commutativity, internalization, fractal dimension) envisaged for it in current physical theory? Or in the face of the complexification and fractalization of the forms of space arising in biology?

Likewise shall we retain the representation of time as a unique parameter in physics or as intrinsically irreversible? And how will the time of physics turn out to be connected with that of biology? By jettisoning its formalization in terms of isolated bifurcations perhaps, in order to take into account the synchronic and diachronic effects in biological systems?

One reason for this rather confused state of affairs (albeit one closer to the often counter-intuitive nature of reality than our spontaneous perceptions can bring us) is that the very epistemological status of space and time remains relatively problematic despite the formal categorization introduced in the foregoing. One additional difficulty has been introduced by the fact that since recent developments in quantum physics, surprisingly mirroring what has long been the situation in biology, we now have to take account of spaces external and internal to the systems under investigation.

We here encounter a distinction long made by philosophers – and notably by Kant – but this time in a form lying on the side of the objects themselves, whereas for Kant it was conceived as lying on the side of the epistemic subject. It is the distinction between space as the form of external and time as the form of internal sense (Kant, 1986b). Such a distinction with respect to objects was unacceptable to the Kant of the Critique, for having broken with an ontological characterization of the objects of natural sciences, any such internality was denied and location was considered something totally external. However, towards the end of his life, in re-examining Newton's *Principia*, Kant could not refrain, on the evidence of the *Opus Postumum*, from thinking through this point afresh, in particular in relation to the question of energy. He renewed his investigation of certain aspects of Leibniz's thought on this subject which he had largely avoided beforehand: he would perhaps have found in the latter-day evidence for a spatialization (and temporalization?) *internal* to the objects of physics and biology, if not the answer to his puzzle, at least a spur to new investigations.

Any attempt to make the internal/external dichotomy correspond in a straightforward way to the distinction between space and time is infected with artificiality – the more so if the distinction is seen as lying within reality itself rather than the knowing subject. Nevertheless, taking into account the Kantian view of space and time as the very conditions for the constitution of objectivity, it does not seem too extreme to speak of them, rather than of subjective "forms of sensible intuition," of objective "forms of sensible manifestation." Such forms could themselves be connected with the concepts of externality and internality. In the case of "the external," one essentially considers the phenomenal manifestation of *relations* between objects (interactions and corresponding measurements). In the case of "the internal" one considers, rather, constraints concurrent with the phenomenal manifestation of the continuing *identity* of an object – or better, of its *identification*.

Notice that the distinction between internal and external aspects is one of the conceptually distinguishing features of biology. But it can be expected to be an element in our conception of the objects of theoretical physics as well, since it now has an objective, mathematically expressed counterpart in the distinction between the external space-time (the base space) and internal spaces (the fibers over the base) which is now a fully developed aspect of the mathematical formalism of key areas of quantum physics. Briefly (and ironically, in the light of certain epistemological tendencies which have sought the reduction of biology to physics), it is on the side of biology that one now looks for the conceptual clarification allowing the development of a more comprehensive abstract framework for the understanding of physical phenomena.

In summary, we have thus arrived at a kind of conceptual "renormalization" (if we may use, by a conceptual extension, this term from physics, mentioned at the end of Chatper 2). The external space-time of physics can be seen as a manifestation of the couple <space/relation> and as providing a "base" relative to which the corresponding couple <time/identifiability> can be viewed as the fibration (the internal spaces). The further articulation of these two elements (via the introduction of supersymmetry and superspace) gives rise to what might be termed a space-time of a kind which allows us to take account of all the aspects in which an object becomes determinate: both its relational properties and its internal structuring activity which permits its continuing identity. In biology, it is the external space-time which corresponds to the manifestation of the <space/relation> couple, while morphogenesis and biorhythms correspond to the manifestation of the <time/identifiability> couple, associated with the identity and functioning of the organism over its lifetime.

All these considerations are partly speculative and demand further detailed elaboration to test their pertinence. On the other hand, the status conferred on our concepts of space and time in the constitution of objectivity has to do with epistemological issues, and here it impacts at a really profound level on the status of categories as basic as causality and on conceptual pairs as important as that of local and global. These categories and concepts are to a greater or lesser degree derived from the intuitive or the theoretical representation of space and time and it is in connection with both intuitive and theoretical considerations that we must give an account of the role of space and time in scientific explanation.

The mathematical organization of scientific theories confers our frame of reference a status and a structure which is objective yet increasingly counter-intuitive: that is to say, basic categories and derived concepts are both increasingly accurate and increasingly unfamiliar in character. We have arrived at a stage where, alongside the development of mathematical conceptual constructions (whether in symbolic or diagrammatic form), *new structural intuitions* are achieved. These intuitions are inherently generated by the mathematical system itself and further and further lacking in directly empirical content.

The task confronting us today is the rational articulation of connections between these new types of (theory-generated) intuition and the experimental results with which we are presented in the realm of physical and biological phenomena. To recall an old distinction from hermeneutics, explanation may make progress, but comprehension may be poorly equipped to follow. We recall Thom remarking, in one of his barbed quips, that quantum mechanics was unintelligible and that everything about it that was rigorous was insignificant. Yet, we have made here an extensive use of ideas coming from quantum mechanics even to indirectly make more intelligible some aspects of the living state of matter.

What is the connection between increasingly abstract spaces and times, theoretically constructed and formally specified, and the intuitively given ones which preside over the development and regulation of our own cognitive capacities? The permanence of our vocabulary, while it constitutes an index of familiarity, certainly does not suffice to justify or explain these relationships. It is to the existence of profound cognitive schemas and invariants of our mental representation and to the way they are transformed that we must appeal. The two epistemic frames proposed (<space, group, equivalence> and <time, semi-group, order>) suggest a way of approaching that difficult question, jointly to the other opposing pairs we examined (local/global, regular/singular ...).

We will return to these conceptual pairs and dualities, in particular in reference to biology, in Chapters 5 and 6. The next chapter mostly goes back to physics, in order to clarify the role of some key (conceptual) construction principles and prepare the grounds for a further comparative exploration.

Chapter 4

Invariances, Symmetries, and Symmetry Breakings

4.1 A Major Structuring Principle of Physics: The Geodesic Principle

In physics, at least in the first stage of rational reconstruction, all seems to be organized according to a fundamental principle which can be found to underlie most formalizations: the *geodesic principle*. Informally, it is a question of minimizing a length (or a surface), given a space and its metrics. We will now explain the variegated accomplishments of this principle according to its field of application. In particular, we will try to present it in its full generality, as a unifying principle of conceptual construction, in physics, in the sense proposed in Chapter 1.

The geodesic principle does indeed appear able to subsume most of the theoretical schemas currently known to physics, not in terms of their specific contents, but rather in terms of the common principles which govern them. It, moreover, has the advantage of being expressible in somewhat ambivalent and clearly bijective terms, according to whether we are relating it to the mathematical structures that it mobilizes or to the great physical referents that it modelizes, thus providing a privileged example of the relationship between mathematics and physics. These relationships are so intimate that they have become core constitutive interactions between and of mathematics and physics. Physically speaking, this principle, which took on various forms throughout history, generalizes the optical principle of the *shortest path* or the dynamic principle of *least action* in the form under which it will find its greatest extension. In fact, provided it undergoes the adequate transformations, it corresponds to the idea according to which any state of equilibrium maximizes or minimizes a correctly selected state function (more specifically: free energy, or partition function, which amounts to the same thing from the statistical mechanics viewpoint ...). All processes take place while optimizing one or several quantities (Lagrangian action, speed of entropy production, ...). As we will discuss in length later on, a determination of this sort singularizes a state or a process among all possible states or trajectories according to specific criteria of criticality (a minimum, a maximum, an optimal path, ...).

A preliminary example: Galileo and inertia Galileo's notion of inertia is an early instance of the geodesic principle, now fully understood within the Hamiltonian frame we will refer to here. It provides by this an example of the audacious speculations that establishing a new construction principle requires. First, it was preceded by confused but extremely deep remarks by Bruno, and then it became a physical principle in the hands of the Tuscan founding father of modern physics. The point is that any empirical evidence suggests that physical bodies do stop if not pushed by a force. Aristotle's observations were perfectly coherent with what we can witness around us. Thus, the sun, in order to move around the Earth, needs to be pulled, at all times, by divine horses or pushed by God's hand or will (St. Thomas). In no circumstances do we see a physical body moving forever, along a vector of constant speed. Yet, by extrapolating from experiments, but daring to push them towards a conceptual limit, Galileo proposed the only pertinent principle, physical inertia. The physical principle constructs knowledge by this sort of "phase transition," as we called it, by which human imagination and symbolic culture go well beyond empirical evidence and propose a unified conceptual frame, at the "limit," in a sense, which makes us understands a large variety of phenomena, independently each of them. In no way is the inertial principle "in the world," yet it beautifully constructs our knowledge as a form of novel objectivity.

The first section of this chapter presents the technical justifications for our approach, which are at times more difficult than the ensuing reflection. These justifications are logically necessary to the reflection, but the autonomy of the other sections (and chapters) should also enable a direct access to the consequences of the great principles of physics that we will evoke, without the reader having to completely assimilate the concepts and notions introduced in this first section. We will therefore first and above all try to present the geodesic principle in its generality, in order to highlight its most relevant characteristics. We will provide more substantial comments and examples later on. The use of numerous technical terms will be unavoidable; we will try to define them as much as possible along the way. The object of this chapter is in fact to present, with a few details, the different aspects and a certain number of applications of this organizing principle of modern physics. For the reader, it would suffice, by means of the albeit incomplete appreciation of this general framework, to see the strength of the geodesic principle: the specific technical details, in particular in section 4.1.1 below, do not constitute a prerequisite for reading the rest of the book (and, at first reading, they may be just grasped in order to get even into the subsequent sections of this chapter).

4.1.1 *The physico-mathematical conceptual frame*

Here are some key hypotheses that underlie the approach we follow.

1) *All fundamental laws of physics are the expression of a geodesic principle applied in the appropriate space.* This is a very general hypothesis, corroborated by the formalisms of the various fields of physics, inasmuch as one is less interested in the particular laws themselves, specific to these fields, than in the principles which govern them and which enable us to derive them mathematically. As we said, it is a question of minimizing a length (or a surface), given a space and its metrics. Mathematically speaking, this signifies putting a calculus of variation into effect. In the field of dynamics, the physics' perspective on the mathematical formulation consists in specifying that it is a Hamiltonian principle applied to an action, given a Lagrangian density.[1] But it could also be, using a more static framework, a question of looking for the minimal surfaces corresponding to an energy minimum for a superficial tension.

2) *The essence of theoretical determination lies upon the identification of the adequate space and of its metric.* Here again, we have the mathematical formulation: the "objects" are supposed to live within a man-

---

[1] Let's recall here that in classical mechanics, Hamilton's principle consists mainly in defining the trajectories as minimizing the Lagrangian action (that is, the time integral of the Lagrangian). In the simplest case, the Hamiltonian, defined in the space of positions and of moments, corresponds to the system's total energy (kinetic plus potential), whereas the Lagrangian, defined in the space of positions and of velocities, corresponds to the difference between these energies (kinetic energy minus potential energy). In field theory the Lagrangian may be obtained as the space integral of the Lagrangian density, which expresses thus the space/local dependency of the Lagrangian. From the mathematical point of view of symplectic geometry, the former resides in the manifold, which is cotangent to the original manifold, while the latter resides in the manifold tangent to this original manifold. In quantum physics, the magnitudes thus represented are not only simple quantities, but operators.

ifold endowed with a metric, and their behavior is determined by extremization of the latter. From a physical point of view, what we have is essentially a precise determination of the system's degrees of freedom and the explicit expression of the Lagrangian density associated with it or, for example in statistical physics of equilibrium, it is the determination of the partition function of a system.

Thus, the characterization of the effective behavior of a system is reduced to the construction of the *relevant space* and of the intended *mathematical function*, which may be roughly made in correspondence with the observables, in physics. Be it a question of phenomenological Lagrangian functions, partition functions, metrics as such, or path integrals (Feynman integrals), the idea of extremization remains the same and effective determination is then nothing but a question of calculi, and is also capable of responding to different methods. What results from taking this principle into account, that is, what are the most profound theoretical and conceptual consequences of this approach? In our view, and based upon its most fecund dynamic interpretation, it is possible to identify the main consequences of which the scope will finally prove to be quite extensive and the gnosiological importance essential for the current intelligibility of the process of construction of objectivity in contemporary physics:

i) The identification of a metric (in the mathematical perspective) or of the Lagrangian density (in physics) enables us to define the transformations of symmetry which leave the equations of movement invariant. The physical aspect is given by the trajectories defined by the Euler-Lagrange equations, that is (mathematical aspect), the corresponding variational equations.[2] It is therefore a question here of characterizing the system's *symmetry group* (as symmetry transformations of a system form a group), which preserves the Euler-Lagrange equations or, more generally, of identifying the gauge transformations (as both local and global transformations – of scale or reference system, say; see the footnote in the premise of Chapter 3).

ii) To these transformations in symmetry correspond, by Noether's theorems (Noether, 1918; Hill, 1951), *invariants* (mathematical aspect) or *conserved quantities* (physical aspect) specific to the system under

---

[2]The mathematical approach may be related to the physical, as the metric element ds^2 may be written as a function of the Lagrangian L, that is as the integral of L over time squared (the metric coefficients g_{ab} are then the potentials). A geodesic thus becomes the minimization of the Lagrangian, as a shortest path in that metric.

consideration and to any systems displaying identical symmetries.[3]

As a consequence, these mathematical invariants (these physical laws of conservation), by extricating the considered systems' most structurally stable quantities, constitute an essential part of their *identity*, that is, that they determine them as mathematical structures associated with the symmetry groups (of invariance) and as physical systems conserving certain quantities during the changes, which affect them. We thus see that the scope of the geodesic principle extends to specifying the essential aspects of the physical system thus considered: what remains invariant during the undergone changes, like a sort of stable base likely to then structure all encountered variabilities and modifications.

As for the role of symmetries, note that many empirical systems are submitted to the same symmetry groups and that the geodesic principle enables us to regroup them under the label of the realization of a same general model. That is, in a way, to categorize them relative to their *fundamental invariants*. One can even go further by showing how these theoretico-conceptual schemes are applied to the theoretical framework of contemporary physics, to relativistic theories (including classical mechanics), quantum theories, and critical theories.

In this perspective, that which fundamentally characterizes relativistic theories, in the more general case of Riemannian metrics, is their *invariance* under the diffeomorphisms of *external* space-time. These are reducible to the Lorentz-Poincaré invariance group[4] and to special relativity, for Minkowskian metrics, and to the Galileo group for the Euclidean met-

---

[3] Noether's theorem is relative to the continuous transformation of symmetries which, operating on the Lagrangians or the Lagrangian densities, preserve their equations of movement such as derived from Hamilton's principle (the Euler-Lagrange equations we have just mentioned). To each of these transformations corresponds conserved physical quantities. Thus, for instance, to the invariance of translation in space (impossibility of physically defining an absolute origin) corresponds the conservation of the kinetic moment; to the invariance of translation in time (impossibility of physically defining an absolute origin for time) corresponds the conservation of energy. From a less classical perspective, to the invariance in a change of the origin of the phases of a wave function corresponds the conservation of the electric charge. These relationships between invariances and conservations will prove to be at the origin of the gauge theories in quantum physics. We will often go back to Noether's theorem.

[4] The Lorentz group corresponds to the rotations in the Minkowski space (of which a component is imaginary); the extension to translations in this space generates the Poincaré group. The Maxwell equations of electromagnetism are invariant under the Lorentz group, this being at the origin of special relativity and of its particular kinetics (contraction of lengths, slowing of clocks, etc.).

rics of classical mechanics. The geodesic principle is directly applicable here to characterize physical trajectories in the gravitational field. By this, general relativity proves to be a theory of gravitation replacing the Newtonian theory which implied absolute space and time, while generating the usual dynamic conservations (energy, momentum, ...). It is even possible to proceed to a unification of this relativistic theory of gravitation with the Maxwellian theory of electromagnetism by means of an increase in dimensions of the reference space followed by a compactification of the additional dimension which we have already mentioned in section 3.1.2, by applying the method of Kaluza and Klein (Duff, 1994) and conserving the geodesic principle within the new space thus defined.

In what concerns quantum theories (quantum field theories), the characteristic invariances are those of gauge groups, Lie groups mainly affecting the *internal* spaces (fibers upon bases of space-time enabling us to specify quantum numbers and the corresponding conservations) (Ryder, 1986). The geodesic principle applies to Lagrangian densities,[5] theoretical or phenomenological, and the gauge invariances appear as the source of the relevant conservation properties. It is under the aegis of such gauge theories and the determination of corresponding invariance groups that the unifications between quantum interactions (electroweak unification, strong unification with strong interaction) can occur. The extremely important role played in these theories by the phenomena of *spontaneous symmetry breaking* will be noted.

Some classical examples of symmetry breaking Let's briefly explain, using a simple example, what we mean by this phenomenon of spontaneous symmetry breaking: a marble placed in unstable equilibrium at the peak of a mound (a configuration presenting a circular symmetry), may at the slightest of fluctuations fall into a state of stable equilibrium at any given point of the base circumference (a configuration now having lost this symmetry of rotation). One will note that each particular drop will form a state of symmetry that is broken relative to the initial state, but if we perform a great number of trials, each direction being equiprobable, the initial symmetry will be restored on average.

Such spontaneous symmetry breaking also occurs during phase transitions (matter's changes of states). For example, a ferromagnet presents, beyond the critical temperature (the Curie point), a null magnetization,

---

[5]The Lagrangian density is the local distribution of energy whose integral over space-time gives the (Lagrangian) action.

and the corresponding magnetic configuration has a spherical symmetry (magnetization is the same in all directions). When we pass below the critical temperature, there appears a magnetic moment vector (a magnetization) favoring one direction in space, the direction of this vector. Likewise, a liquid such as water presents omnidirectional symmetry properties. On the other hand, if the liquid is cooled down below the freezing point, crystals will appear, breaking this symmetry to the benefit of another, a lesser symmetry, that of the corresponding crystalline structures.

In quantum physics, symmetry breaking is supposed to generate the mass spectrum of the interacting particles, *via* the interactions with the Higgs bosonic interaction field (which has not yet been experimentally demonstrated). And numerous physicists also attribute the birth of our universe to a highly energetic spontaneous symmetry breaking of the "quantum vacuum," a hypothetical initial state ("Big Bang" theory).

These breakings of symmetry constitute also the most spectacular manifestation of critical phenomena and of the corresponding critical theories (phase transitions). From a thermodynamic point of view, the geodesic principle here takes the form of a characterization of equilibrium by the minimization of the system's free energy, or the form of an expression of a criterion for irreversible evolution outside of equilibrium by extremization of the speed of entropy production (Prigogine and Stengers, 1988; Nicolis, 1986; Nicolis and Prigogine, 1989). From the view point of statistical physics, as we have already mentioned, it is the system's partition function which presents these properties of extremality. During critical transitions, the dominant physical invariance is namely an invariance by dilatation in that the correlation lengths become infinite, enabling us to put into motion the renormalization procedure (already used in the context of quantum field theory to eliminate certain divergences, as mentioned in Chapter 2). This symmetry breaking has recourse to the application of the renormalization (semi-)group. But another, more abstract type of invariance then appears, this time mainly concerning the behaviors of the critical exponents themselves, in relationship to the existence of "universality classes," which discard the specificities of particular systems to only retain characteristics that are as general as the dimensionalities of reference spaces (be it the spaces of definition of the order parameters or the embedding spaces).

Examining these examples, we can thus see that the recognition and application of the geodesic principle in fact opens a problem which appears even more essential, that of the symmetries and breakings of symmetries a physical system may present or possess inherently. But before moving on to those more general considerations relating to symmetries and mathematical invariances of physical systems, it appears indispensable to attempt to characterize the gnosiological importance of this geodesic principle. In this regard, we have already stressed its exceptional outreach in the formal determination of the objectivities and its organizing power for scientific intelligibility, but these aspects do not appear limited to purely technical reasons and seem to extend beyond the field of physics, appearing as an invariant of human cognition itself. This is easily conceivable if we consider the fact, evoked above, that a geodesic characteristic, regardless of the concerned reference space, is also a *critical* characteristic: a trajectory or a quantity which is minimized or maximized.

Now all possible trajectories or all possible measurements present generic traits in contrast to specific or critical ones. The distinctions between them should be clear: generic trajectories or evolutions are the many, indistinguishable and possible ones, and differ from the specificity of critical paths, which, by this fact, find themselves to be selected "spontaneously," being immediately distinguishable (a key property of geodesics). It very well appears that from the cognitive viewpoint, within the immensity of genericity, only profoundly specific (in particular geodesic or critical) traits are able to operate as elements of objective determinations. These range from the role of optimal pathways for action, from the practical identification of minimal trajectories for the body's movement, to the topological catastrophes associated with the borders, which are apparent within visual perception. This is also what must have led us not only to seek and formulate physical principles, but also, unconsciously, to select geometric frameworks of reference stemming indeed from experience, but referring at the same time to this specific characteristic, such as classical Euclidean geometry, or leading to a search for minimal axiomatics in the characterization of formal systems. Note that Euclidean geometry is a "critical case" of Riemann's approach: its spaces have zero curvature.

Geodesics or critical paths, both in physics and in biology, currently make interesting approaches to complexity based on what some do not hesitate to call the "landscape paradigm" (Sherrington, 1997). The landscapes in question correspond to multidimensional surfaces (of energetic or other nature) and the behaviors of the systems, which they represent, are governed by the evolution of parameters within this landscape with regard to its "passes," "valleys," "peaks," etc., that is, in fact, its set of critical points or zones. The geodesic principle would therefore not just play an essential role in the formal determination of the physical objectivities; it would at the same time meet a very profound cognitive constraint of specific discernability upon generic "backgrounds," the discernability corresponding to a critical characterization.

The geodesic principle and its consequences are thus highlighted in the essential role of continuous symmetries and of Noether's theorem in the determination of physical systems and of their properties. We may attempt to go further in our investigation of the theoretical place occupied by symmetries by including, alongside these continuous symmetries, the examination of the discrete symmetries which can be encountered in physics (and of their breaking), thus highlighting the essential epistemological importance acquired in this field by these questions. Indeed, the existence of discrete symmetries may constrain certain physical properties just as strongly. It could be an issue of abstract symmetries: for example, the symmetry or anti-symmetry properties under permutation of the elements presented by the wave functions of assemblages of quanta govern their collective behaviors, that is, the nature of the statistics to which they obey: Fermi–Dirac statistics and the Pauli fermion exclusion principle (principles of impenetrability and of spatial extensionality of "matter") for antisymmetry, Bose–Einstein statistics and Bose condensation of bosons (properties of superposition of fields of interaction) for symmetry. But it may also be a question of symmetries touching upon the properties of the matter-space-time complex itself (Cohen-Tannoudji and Spiro, 1986). Such is also the case with the symmetries under the operators of parity (P) (inversion of spatial directions), of time reversal (T), or of charge conjugation (C) (passage from matter to antimatter). The CPT theorem spectacularly associates these three operations: it states the general invariance of the laws of physics under the composition of the three operators, currently experimentally verified in a very precise way. We know that symmetry breaking associated with these operations involve questions as profound as those concerning the intrinsic chirality (right-left orientation) of reference systems (breaking of P within weak interaction – disintegration for instance), the prevalence of matter over antimatter in our universe (Sakharov, 1982), or even the intrinsic time irreversibility of the phenomena (breaking of CP, highlighted by the disintegration of a K meson into two pions, for instance). Let's emphasize the fact that, despite the quite surprising aspect of such a result and despite the important eventually derivable cosmological consequences, this possibility of the violation of CP already exists within the framework of the contemporary so-called "standard" theory, at the level of the distribution of quarks into distinct families (there needs to be at least three families for this possibility to arise and the current standard model has exactly three).

Thus, be they continuous or discrete, symmetries take an important place within the analysis and interpretation of contemporary physics. And, by Noether's theorem, they are related to conservation properties and, thus, to geodesics. It is therefore legitimate to proceed to an even more attentive examination not only of their theoretical role, but also of their cognitive and even foundational roles, along Weyl's views, quoted in Chapter 1.

4.2 On the Role of Symmetries and of Their Breakings: From Description to Determination

During the preceding discussion, we have seen the essential role, beyond the very expression of the geodesic principle, played by the properties of mathematical symmetries and their breaking for the characterization of the most fundamental of physical systems. In fact, it very well appears that, in the analysis of their consideration as well as in the determining role they are called to play, we have progressively gone from the description of the essential role which symmetries technically play within physical theories to an attempt to interpret the way in which they organize the knowledge we have of reality, or even the way in which they determine this physical reality itself by contributing to constitute it. It is from this standpoint that we will attempt to continue our investigation pertaining to the most fundamental objective determinations regarding the natural sciences. To do so, we will successively examine the relationships that symmetries and their breaking maintain with profound logical structures, as well as to examine a possible analysis of their role in the satisfaction of certain conditions of possibility for objective knowledge and thus of their role in the determination of scientific reality. In short, symmetries organize knowledge and its friction over "reality"; they actually organize our world of intelligible action.

4.2.1 *Symmetries, symmetry breaking, and logic*

4.2.1.1 *Symmetries, group structure, and logical equivalence relations*

If we may wholeheartedly admit the existence of relationships between the structures of perception or even of intellectual cognition and the properties of symmetry or symmetry breaking affecting the objects under examination, there arises the opposite question of which relationships may these properties maintain with logic *stricto sensu*. What we claim here is not that logic contributes to structuring these cognitive capacities in a more or less coherent and *a priori* way, but rather that it constitutes itself as a formal discipline correlated to the constitution of mathematical concepts, in their manifolded groundings. And this since ancient Greece.

We know that the mathematical group structure constitutes the most adequate means for the study and treatment of symmetries and that it plays an essential role in the characterization of the invariances of a system, to an extent where some authors have made it into a fundamental element of all scientific knowledge (Ullmo, 1969). Let's briefly demonstrate how this mathematical structure is itself in relationship with a more strictly logical structure, the *equivalence relation*. Let's recall that a relation R (between two terms x and y) is said to be an equivalence relation if it is:

i) Reflexive (xRx);
ii) Symmetrical (xRy) implies (yRx);
iii) Transitive (xRy) and (yRz) imply (xRz).

Moreover, it is demonstrated that an equivalence relation presents the property of partitioning the set of elements to which it is applied into disjointed subsets of which the reunion reconstitutes the whole set. Now, with a few minor adaptations, we may put this logical structure into systematic correspondence with the mathematical group structure (in the following, A, B, and C are the group's operations and E designates its neutral element; a, b, c, d are objects upon which these operations are performed):

i) reflexivity can be put into correspondence with the existence of the neutral element E of which the application upon any object leaves the object unchanged (E exists such that Ea = a)
ii) symmetry can be put into correspondence with the existence of the inverse operation of which the application on an element causes a return to the original element (for a given A there exists A^{-1} such that if $Aa = b$ then $a = A^{-1}b$)
iii) transitivity can be put into correspondence with the property according to which the product of two of the group's operations is still an operation of the group (there exists C such that if $BAa = b$, then $Ca = b$, that is, $C = BA$)
iv) as for the associative property of the operation products required by the axiomatic of group structure, it is put into correspondence with the

simple result, associated with the other properties of the equivalence relation, according to which [if xRy and yRz and zRu, then xRz and zRu (application of transitivity) and the second (for the same reason) reduces to xRy and yRu; both are finally identified as xRu].

Thus, by this mapping of the respective axioms of existence, we highlight the close formal relationship between the equivalence relation and the group structure. We may even assert that the mathematical structure of an abstract group constitutes a privileged, non-trivial instantiation of the logical equivalence relation. Van Fraassen (1989) goes even further when asserting that "*the three concepts* of equivalence relation, of partition and of transformation group *actually amount to a single concept.*" As a matter of fact, we demonstrate that, as the equivalence classes of an equivalence relation form a partition, there exists a group of which the invariant sets are the partition's cells and that each group defines the equivalence relation: "to be transformed one into the other" for the elements of the group. Thus, as far as we are concerned, instead of speaking of a single concept applicable to various fields, we prefer to consider such a situation as a very strong formal correspondence, an isomorphism, between distinct concepts, each being adapted to its own disciplinary field (sets for mathematics and equivalence relations for logic). Because the important thing is to be able to construct such a correspondence between the structure of "sensible" symmetries, the "intelligible" group structure and the "logical" equivalence relation, thus somewhat comforting the point of view according to which identification and operational use of symmetries stem from very profound cognitive acts. Moreover, we have considered that an essential part of the "identity" of an object of study could be condensed to its symmetry properties (the other part being mainly an issue of symmetry breaking), it is thus constituted a sort of epistemological complex which goes from the object's characterization linked to the cognitive capacities of perception to that which is the most formal, represented by the logic of the equivalence relation, and passing by the determination of the relevant symmetries and the mathematical group theory.

4.2.1.2 *Symmetry breaking, semi-group structure, and order relations*

But if we take the path of the comparison between the mathematical group structure – an abstract theory of symmetry properties – and the logical structure of the equivalence relation, it is then legitimate to question ourselves regarding the possibility of finding similar elements of comparison for the phenomena associated with symmetry breaking or at least associated with some of them. In order to attempt to respond to this, we may notice that while considering earlier the theory of critical phenomena and the associated symmetry breaking, we had evoked the importance (for their treatment and the determination of new invariants in terms of universality classes) of the process of renormalization (Chapter 2, section 2.2.2). Now the latter appeared to be indissociable from the mathematical structure of the semi-group, which we know to be differentiated from the group structure only by the absence of inverses of the operations which constitute it. It is then tempting to relate this mathematical structure of the semi-group to the logical structure of the *order relation* which, for its part, differentiates itself from the equivalence relation only by the absence of the symmetry axiom (the (ii) of the equivalence relation), that is, that which we have put into correspondence with the existence of the inverse operation for the group axioms. This is how the filiation would be constituted: symmetry breaking, renormalization semi-group, order relation. As the counterpart of the preceding sequence (symmetry, group, equivalence relation), this frame constitutes a rather coherent portrait of the sought correspondences. Moreover, it relates to our previous analysis of the asymmetry of relational, irreversible time (the "time arrow").

4.2.2 *Symmetries, symmetry breaking, and determination of physical reality*

We have just been considering a somewhat global approach, founded firstly upon the generalized usage of the geodesic principle within the framework theories of physics, to the conceptual and theoretical landscape which seems to have progressively organized itself around the notions of symmetry and of symmetry breaking from increasingly abstract constructions. It thus appears that these concepts constitute another breakthrough in theoretical generality to a point where, beyond their operational use, some epistemic principles find themselves put into perspective. Let's begin addressing this point by providing a few examples.

Indeed, a first resulting aspect concerns the way in which the concept of *causality* is shaken, especially when one attempts to relate it to the concept of force. Thus, general relativity highlights the duality existing between the characterization of the geometry of the universe and that of energy-momentum within this universe. By this duality and by the effects of the invariance principle under the diffeomorphisms of space-time, the "forces" are relativized to the nature of this geometry: they will even appear or disappear according to the geometrical nature of the universe chosen to describe the physical behavior. Likewise for gauge theories, concerning the time internal variables: as in the case of relativity, the choice of gauges and of their changes enables us to define certain interactions, or, conversely, to make them disappear. If we consider that one of the modalities of expression and of observation of the causal processes is to be found within the characterization of the forces and fields which "cause" the observed phenomena, we will see that this modality is thrown into question by the effects of these transformations. Not that the causal structure itself is jeopardized, but that the description of its effects is relativized. This conduces to formulate a representation of it which is more elaborate and more mathematized than the representation resulting from the intuition stemming from classical behaviors. We have already synthesized this aspect in Chapter 1 and recall it here because it is central to our approach: causes become interactions and these interactions themselves constitute the fabric of the universe of their manifestations; reshaping this fabric seems to modify the interactions, intervening upon the interactions appears to reshape the fabric. We will also return to this in the next chapter.

Moreover, concerning the concept of *observability*, we have seen that the existence of invariances under symmetry operations is equivalent to the expression of formal restrictions to certain possibilities of observation, due to the absence of absolute origins and reference systems for space and time. Thus, gauge theories and relativity groups set the limits to the knowledge we can obtain about a system and to the quantities we can measure.

What is amazing is that it is precisely these principles of relativity, these limits imposed upon observation, which contribute to determine, in a very strong sense, the physical objectivities which conform to them. These are exactly constituted by invariance properties with respect to non-absolute reference systems. An analysis of the previously mentioned examples demonstrates that an object of fundamental observabilities is always relative to the *framework of reference* to which the phenomena are put into relationship: the space-times of various natures (including highly abstract spaces) within which they are defined. Thus, the *determination* of physical objectivities seems correlative to the (regulated) *indetermination* of the reference systems relative to which they manifest themselves. As Weyl puts it: the constructions of objectivity begin by choosing a reference system and a measure in it; then, by the analysis of the resulting invariances with respect to the reference systems.

More fundamentally still, it appears that the symmetry/symmetry breaking pair is beginning to play, for the intelligibility of physics, a role similar to the one already played in other disciplinary fields by certain foundational dichotomies such as, in mathematics, the dichotomies of finite/infinite, continuous/discrete, local/global (Lautman, 1977). As we observed, the dialectic of the symmetry/breaking of symmetry pair thematizes, on the one hand, invariance, conservation, regularity, and equivalence, and on the other, criticality, instability, singularity, and ordering.

We have seen that through the pair's dialectic, it is an essential component of the very *identity* of the scientific object that is presented and objectivized. Could we go even further and consider that we have thus managed to construct this identity at a level such that cognitive schemas conceived as conditions of possibility for any construction of objectivity are henceforth mobilized, thus reviving a form of transcendental approach (Petitot, 1991)? In fact, we showed that there exists a close formal relationship between the abstract properties of symmetry captured by mathematical group structures and logical structures as fundamental as the equivalence relation. At the same time, there may exist a similar formal relationship between the semi-group structure, of which an essential usage is made with regards to the renormalized treatment of critical phenomena, and the logical structure of the order relation. Now, the theoretical analysis of the abstract notions of *space* and of *time* demonstrates that, for their formal reconstruction, these notions need to mobilize the mathematical structures of group and of semi-group, respectively. Indeed, regardless of the number of dimensions considered, the displacement properties, consubstantial to the concept of space, refer to the determinations of the displacement group, whereas the properties of irreversibility and of the passing of time refer to the characteristics of the semi-group (generally, for one parameter). We then witness the constitution of a pair of abstract complexes which doubtlessly represents one of the essential bases for any objective interpretation within the processes of the construction of knowledge, as we have mentioned earlier: the complexes of <space, group structure, equivalence relation> on the one hand and of <time, semi-group structure, order relation> on the other. Let's point out once more, in order to dispel any possible confusion, that the space and time evoked by these complexes no longer refer to physical entities as such, but rather to the conceptual frameworks, which are meant to enable any physics to manifest itself, that is, to abstract conditions of possibility and not to effective realizations, thus reactualizing a point of view (Kant, 1986a). That is, space and time are no longer considered as "objects" to be studied, but rather as the conditions of possibility for any sensible experience. In this sense, the symmetries and breakings of symmetries associated with these gnosiological complexes appear not only as elements of the intelligibility of physical reality, but indeed as factors for the scientific constitution of such reality.

So, could we say that we have clearly passed from a descriptive analysis of the theoretico-conceptual situation currently dominant within the epistemology of physics to that which may constitute the quasi prescriptive foundations of a contemporary "natural philosophy," where the constitutive principles are to be found not only in the use of mathematical structures, but also in the formal redetermination of these "forms of intuition," which are space and time? Not only would we simply operationalize them, but by coupling them with the corresponding logical and mathematical determinations (group structure, equivalence relation, etc.) we refer them to the frameworks of invariance, which make them into reference structures that are mathematically determined, rather than abstract and vague.

The concept of *invariance*, having already demonstrated its tremendous technical and theoretical fecundity in mathematics and physics, now appears to be called to play an essential role in both the understanding of cognitive processes themselves as well as in the formal reconstruction of their conditions of possibility, doing so with and in contrast to the concept of *critical transition*, by its invariance and symmetry breakings. Husserl had demonstrated this on a philosophical level, following all the mathematical advances of his time, and Thom semiotically remathematized it with the application of his "catastrophe theory" (critical situations manifest as singularities). And this is what we are trying to readdress when we refer to the fact that mathematics constitutes the field where concepts and demonstrations are, epistemologically speaking, the most structurally stable (invariant!) among all forms of knowledge, all the while analyzing the constitutive role of mathematics for a great number of scientific disciplines (it is by this a singularity within our symbolic culture). But there are still branches of the natural sciences, which, while conforming to scientific method and rigor and having recourse to their own formalisms, are still not mathematized in the way that physics can be. They continue, however, to play a more descriptive than prescriptive role, one that is more classificatory than predictive, in that they are not very "calculatory." Such is the case with biology where keen questions relating to invariance and variability are posed. It therefore seems interesting and instructive to try to test this general approach using this particular case in order to analyze its scope and limits, albeit summarily.

4.3 Invariance and Variability in Biology

4.3.1 *A few abstract invariances in biology: Homology, analogy, allometry*

In biology, there are numerous examples of symmetry properties and of sequences of symmetry breaking, be it an issue of morphology, of functionality, or of embryogenesis and development. But rather than sticking to a dominantly descriptive approach to such properties that are lengthily exposed in several texts (see for instance Vogel and Angermann, 1984), it appears conceptually more interesting to try to fully enter into the very foundations of modelizations and to consider more general principles, concerning both the apparition and selection of several great levels of organization as well as the transformation of organic structures and of physiological functions (in this regard, see also the works of Bouligand, 1980, 1989; Dellatre and Thellier, 1980; Kauffman, 1993; Thom, 1972, 1980). It is this aspect which we would attempt to briefly explore here, in order to highlight some of its important characteristics, namely within the framework of the theory of evolution and of the interpretation that we may provide for some of its realizations.

Indeed, if we consider not only specific organs or organisms, or the characteristics of specific species, but if from the onset we consider a more global viewpoint such as the comparison between different species, be it at a specific point in time – synchronously – or within an evolutive process – diachronically –, we will be led to consider other types of invariances and of symmetries than those which we find morphologically or functionally represented within specific biological structures. From an evolutive standpoint, these are mainly of two types: those resulting from the evolution of the same fundamental organ towards organs which may be quite different but which have a common origin, therefore entertaining relationships of *homology*, and those which, by convergence, result from the transformation, under the pressure of selection, of fundamentally different organs towards similar forms and identical structures that entertain therefore relationships of *analogy*. From a more immediate point of view and, geometrically speaking, one which is more metrical, we would have for comparable biological structures relationships of *allometry*, that is, in the last analysis, scaling laws corresponding to the invariances of certain laws by dilatation (see Chapter 3 and below for the notion of scaling and allometry).

To begin, let's consider homologies (same origins, but often having different functions) and analogies (same functions, but different origins). They are to be found throughout all living forms. In plants for instance, where tubercles, tendrils, and stem thorns are homologous and where roots, stems, and tendrilled leaves are analogous. In animals, the articulation of the limbs of vertebrates can be found homologously in the fins of the dolphin, in the wings of flying reptiles and of birds or bats: paleontology shows us a common origin, despite their differing functions. The mouthparts of insects are also homologous, with their rather variegated functions, and so are the cerebrums of vertebrates. While analogies are discernible between the different types of excretory organs, between the wings of birds and of insects, between burying and shovel-shaped legs of the mole and of the mole cricket, or even between the tubercle of the potato and that of the dahlia (in this case the analogy is morphological).

These comparisons and classifications are not limited to the relating of organs and functions, they also extend to behaviors. Thus, there exist homologous movements among different species of duck, and as well there often exists for a number of other species a sequence of homologous behaviors which derive from an original specialized typical behavior. Likewise, many gestures, stemming from situational convergences, are analogous between numerous species, such as acts of appeasement in the face of the threat of aggression. With regard to the human species, the comparison with animals highlights similarities and differences: as for the other species, there exist homologies associated with acquired phylogenetic traits stored within the genes, as well as analogies associated with acquired ontogenetic characteristics stored within the central nervous system. However, there also exist traditional homologies where memory replaces the genotype as the locus of information storage. Despite its exceptional character, it remains necessary to note that this type of traditional homology, transmitted via the memory of a social group, does not appear to be completely lacking among other animal species, such as demonstrated by the behavior of the Japanese stumptail macaque or by the song of the African pin-tailed whydah, and many other animal populations, which have been shown to have culturally transmitted behavior. All these characteristics are of interest to us inasmuch as they manifest invariances which are interspecific, organic, functional, and behavioral, beyond the variability associated with the evolution and differentiation of life forms, a little bit in the way that geometrical or more abstract symmetries display and even prescribe invariances in the variety and the apparent dispersion of physical phenomena (whereas in biology, we remain at a stage which is essentially descriptive).

With the properties of allometry (Thompson, 1961; Gould, 1966; West et al., 1997) and of scaling (Peters, 1983; Schmidt-Nielsen, 1984), we are no longer within a context of invariances associated with an original form, organ, or behavior type. Their homologous traces are to be found in the evolutive variations or in the permanence of functions which, by the selective pressure they exercise, govern the convergence of the forms or of the analogous behavior. Now, instead, at a specific time in the history of evolution, and still within the framework of interspecific comparisons, it is a question of identifying and interpreting "scaling invariances," expressed by the scaling laws already mentioned in Chapter 3. These laws may be witnessed between elements which are common to these species and relevant to their characterization (metabolisms, lifespans, internal rhythms, ..., but also relationships between physiological systems such as the nervous, respiratory, and vascular systems). In a very general way, these allometric relationships are expressed in power laws, as we have seen, where the variable is the characteristic size of the species under consideration, most often measured by means of the adult's average weight, M, and is therefore of the form:

$$Q = AM^k.$$

In this expression, Q represents the quantity under consideration, for a species of which the average adult weight is M. A is an interspecific constant which is characteristic of this quantity and k the scaling exponent, which is also interspecific as well as being characteristic of the quantity at hand. Therefore, in principle, relative to such quantities, a whole set of species can find themselves grouped under a single class characterized by the series of these invariable quantities which are the magnitudes A and k. The distinctions only appear when the parameter M is taken into account, distributing Q into different sizes, each specific to a species.

Let's consider a few particularly significant examples. All metabolic rates scale as the 3/4 power of mass (see Chapter 3), thus indicating a strong structural stability for living organisms in the mobilization of the means of survival. Likewise, regardless of the species, all characteristic biological time spans (lifespan, gestation period, maturation period, inverses of breathing frequencies, heart rates – when there is a heart) appear all to scale at the same power of mass, $k = 1/4$. This characteristic has rather interesting consequences in the cases of periodic physiological phenomena. Particularly, let's recall that mammals (to which correspond, for a given physiological function and their lifespan, a single constant A, with the same value) are characterized by the same total number of breaths (2.5×10^8 approximately) or of heart beats (10^9 approximately), whereas the corresponding frequencies or periods are obviously different for each species. Thus, the passing of the singularity from the specific to the general is accompanied here by the passing from dimensional magnitudes to pure numbers that are characteristic of the generality thus constructed, a manifest consequence of the invariance thus revealed.

Let's just mention that there exist scaling properties, which lead to allometric relationships other than the interspecies and mass related properties we have just been considering. For instance, this occurs within the same species, as in the case of the rat for example, over the course of its development, over its period of growth (Burri *et al.*, 1977), for the ratio between the total pulmonary or alveolar capillary surface and the lung's volume, or as in the case of the dog, for the post natal growth of its bronchial tree (Horsfield, 1977). But in these cases, and conversely to those which we have examined precedently for the interspecies relationships, we cannot identify manifest invariances. Therefore, the resulting symmetries by dilation remain confined to a specific morphological description, even if we can relate them to morphogenetic dynamics which enable us to account for their development (Bailly *et al.* 1994).

Let's now return to the principles, which can eventually be derived from these observations and analyses: even if we observe that redundancies and degenerations are extremely frequent in biology, it very well appears that certain symmetries conform to principles if not of economy, then at least of *optimization*, as demonstrated by the analysis we can make of the existence of fractal geometrical structures among many organs engaged in vital physiological functions and which must, let's recall, ensure the compatibility between three necessities which may appear contradictory: to ensure maximal efficiency and flexibility of control (thus the forming of very large active interfaces or of arborescent networks of which the surface of the terminal section is very high in order to involve a whole volume), while presenting only an acceptable level of steric encumberment (if not minimal volumes, then at lest very limited ones), and behaving as truly differentiated organs. These organs, typically the lungs or the vascular system, present within this determined spatiality a homogeneity of structure and of functioning which individuates them in a non-ambiguous fashion. One will note here the formal affinity of this quest for optimization, which we have just mentioned, with the rules of optimization that are concomitant to the geodesic principles of physics discussed in length earlier.

In other cases, it appears, for instance in what concerns the symmetries of temporal periodicity, but also in spatial arrangements, that principles of *synchronization* (spatiotemporal, then) are manifested, enabling the regulation and integration of differentiated structures and distinct levels of organization within a whole which is sufficiently perennial for survival and reproduction. These effects presumably derive from the evolutionary history of species and a number of them must be associated with the observed homologies, which translate, as we have seen, an invariance of which the origin is to be found within the existence of original forms to derived forms.

These principles are clearly not arbitrary, and must stem from evolutive responses in terms of adaptation to environmental or functional selection pressures. Therefore, they cannot generally be dissociated from the conditions of apparition of the structures to which they apply and from the transformations to which they have been submitted due to the historic modifications of the environments to which they have been exposed. However, in all likelihood, they also stem from internal tendencies that are more local but nonetheless important, by the fact that living systems also present at some levels physico-chemical aspects, which are governed by the laws and principles of physics, as local modelizations, by means of non-linear dynamic systems clearly show.

4.3.2 *Comments regarding the relationships between invariances and the conditions of possibility for life*

After these few discussions regarding certain symmetries and regularities encountered in biology and the regulating principles to which they could refer, we would like to orient the reflection towards the specificity of the biological with regard to the physical in terms of invariances, symmetries, and of the epistemological meanings we can derive from them.

If we attempt to outline a rough classification of the set of determinations involved in living phenomena, we would recall that any living organism is characterized from a *topological* point of view by the distinction between an inside and an outside, thus enabling us to define a spatial individuation. It is also characterized, from an *energetic* point of view, by the existence of a metabolism regulating its exchanges by fluxes between inside and outside, as well as, from an *informational* point of view, by a genetic expression morphologically and functionally translated by a structuration into levels of organization[6] ensuring both regulation (mainly top-down functions) and integration (mainly bottom-up oriented activities). Finally, it is characterized from an *"identity"* point of view by a reproductive capacity corresponding to the transmission of this genetic and other heritages and of its aptitude for evolutionary adaptation. We must also mention this extraordinary invariance, throughout time and between different species, which constitutes the stable universality of the genetic structure itself. These aspects may or may not need to be applied to the field of physics: the inside/outside distinction is in no way necessary to the characterization of a physical object, even if these determinations may prove to be relevant in defining physical systems. Conversely, the issue of energy exchanges is common although strongly qualified by the fact that biological individuation has no bearing on it, etc. If, furthermore, we are interested in the semi-empirical conditions of possibility for the existence of such biological objects, we will observe, as we have already indicated in Chapter 3 and further specify below, that from a spatial point of view, it very well appears that the embedding space (or, explicitly, the space within which life manifests itself) must be three-dimensional. From a temporal point of view, it will be necessary to articulate very distinct types of temporality and, from the intrinsic and abstract structural point of view, a functional hierarchy of levels of organization will need to be constituted. We shall not return to this last point, which we considered earlier under a somewhat different angle. However, we shall continue here with a more detailed view upon an idea outlined in the preceding chapter concerning the issue of spatial dimensionality.

Some (meta)physics at few dimensions. Let's more expand on a discussion already mentioned. Could one conceive of a biological science which would not constitute itself within a three-dimensional embedding space (in the way that we have uni- or bi-dimensional physics, for instance)? This question refers to the organizational constraints and the necessity of articulating the local and the global, the whole and the parts. Indeed, it is

---

[6]We use here the term "level of organization" in a strong and non-metaphoric sense, responding to a double determination: the change in level of organization must be accompanied on the one hand by the divergence (passing to the infinite boundary) of an intensive magnitude contributing to characterize the original level. On the other hand, it must be accompanied by a change in the relevant object considered by the theory or experiment (a way to signify that from this viewpoint the extensional composition of elements at one level does not suffice to characterize both the nature and the behavior of the object considered at a superior level). We will return to this in Chapter 6.

technically necessary to couple a global (individuating) homogeneity with local distinctions, in a way that is not only compatible, but also cooperative, and in order to ensure high levels of connexity all the while avoiding confusions and mix-ups. This can lead to the following *a posteriori* reasoning:

1. If the number of spatial dimensions (d) is less than three, we will lack homogeneity and connexity, because on the one hand, neighborhoods between elements are too few (four closest neighbors within a square network, for instance) and, moreover, the independent routes of communication are too restricted or too complicated and have the risk of intersecting and mixing; the situation may be considered as non functional, by excess of differentiation and by necessity of a too high level of structural complexity. In short, crossing channels would intersect and mix.
2. If, conversely, $d > 3$, then the situation will be the opposite: the levels of connexity (many channels are possible) and homogeneity are too great to preserve a sufficient local differentiation having recourse to elaborate structures governing transport and exchange, ensuring the global functions and regulating the local situations; for instance, in physics, mean field theories are exact from $d = 4$ and up.[7] This may be interpreted by saying that there arises no necessity for distinction and discrimination between structures in order to ensure a globality from local interactions. The situation is this time non functional by an excess of homogeneity and lack of structural differentiation.
3. Finally, if $d = 3$, we can then hope to combine a global connexity that is just sufficient to coordinate an enormously differentiating heterogeneity, albeit by having recourse, for the articulation between structures and functions, to these somewhat particular geometries, the fractal geometries we have just mentioned.

Physics thus provides good reasons for mathematics to propose three dimensions for the realm of the biological. The reader may take this as he/she likes and consider that, in a way, we have answered the metaphysical question of *why* we live within a three-dimensional space.

The preceding remarks and investigations have thus led us to note that the terms of symmetry and of invariances take on meanings in biology,

---

[7]In short, according to these theories and their various contexts of applications, the (observable) values at one point are determined, as averages, by neighborhood points' values.

which are akin to those prevailing in physics, without, nevertheless, being reduced to them. Indeed, if these properties are frequently referred to and expressed within our space-time (geometrical aspects, temporal periodicities), they correspond just as often to much more abstract principles inducing a recourse to internal spaces, which, in contrast to that which occurs within the abstract internal spaces of physics, remain largely underdetermined. Such is the case, for instance, with the symmetry by invariance of the species' genetic heritage from one generation to another despite its individual variations, despite the concomitant invariance of great levels of organization or of functional systems, or even the invariance of individual or social behaviors as demonstrated by ethological studies. Complementarily, we have noted several times the importance of symmetry breaking for biology, be it a question of contrasting the differentiated biological activity (metabolic, enzymatic, ...) of molecules having left or right configurations (sugars, proteins, ...) or, at a very different level of organization, be it an issue of the irreversible character of the development and aging of organisms. To a point where many have wondered about the intrinsic relationship between the existence of life and that of certain fundamental symmetry breaking. These may be sometimes of a spatial nature (in relationship to differentiation and the organization into different levels, for example during embryogenesis), but are mainly temporal (at the individual scale – development then aging – as at the global scale – evolution).

Finally, and to conclude here with regard to the field of biology, we would like to return more specifically to the possible relationships between physics and biology, by taking a point of view which is completely formal and beyond their respective theoretical and conceptual contents, the latter being our habitual standpoint. By attempting to place ourselves at a point from which we would seek to mutually posit physics and biology, and by now conceiving each of these disciplines as a *model* of natural science, almost in the *semantic* sense of logical model theory. The logico-formal exercise may propose an ulterior view upon our monism, which feeds on theoretical differentiations all the while seeking possible unifications between theories. The reader will be able to follow it even without knowing the terminology of logic.

To do this, let's suppose we have a general axiomatic framework for the purpose of generally characterizing the scientific rationality of the natural sciences and its formal and methodological requirements, even if they are thematized differently according to discipline and sub-discipline. Now let's suppose that physics and biology constitute models of realization, still in the sense of model theory, of this possible axiomatic. Let's try to specify from this point of view their relationships, considering that through one of their central problematic concepts, that of *matter*, they will meet, partially at least, within the same semantic domain. In which sense could we say that physics constitutes a standard model of this axiomatic whereas biology presents a non-standard model?

From the semantic point of view, it would be tempting to consider biology as an "enlargement" of the physical model, that is, by adding more "structures" (more organization, more observables ...). This may be roughly assimilated with the technical sense given to the notion of enlargement or extension in logic and model theory. This enlargement would enable us to distinguish the concept of *life* within matter, a concept and a reality that physics does not know how to "see" nor to "name." Syntactically (*cf.* the IST approach of Nelson (1977)), we would then consider that the introduction of a new predicate ("*non-living*") enables us to formally distinguish purely physical situations from biological ones. This predicate would not be expressible within the theory, in the same way that formal arithmetic cannot separate the "standard" number from the "non-standard" by means of an internal predicate.

Such an approach, which has nothing of the absoluteness specific to arithmetic or to formal set theory, only suggests a possible conceptual framework for the unity and diversity we confer to natural, physical vs living phenomena. From an epistemological or even philosophical point of view, it enables us to choose between the terms of an alternative:

- either, as it is often admitted, the axiomatic (virtual) presuppositions which govern the constitution and development of physics and of biology are fundamentally different, and these disciplines develop non equivalent models from this point of view; we would then be facing an underlying dualism, within which the concept of "life" not only plays a discriminating role, but also corresponds to an axiomatic irreducibility (we could even find, theoretically, biological finalism, or even vitalism);
- or, according to the proposition we have just presented, we would consider a single axiomatic for both disciplines, which would refer to a sort of formal materialistic monism, which is nonetheless likely to lead to the realization of semantically distinct models (physics and biology being distributed between standard and non-standard, respectively); syntactic analysis reestablishes a duality in this monism, but the irreducibility is no longer axiomatic, it refers to the distinction between

internal and external sets. This would be compatible with our view of life as a "physical" singularity, an *external* or non-standard boundary case, having a more specific counterpart the notion of an "extended critical situation" to which we will return in Chapter 6. Strict theoretical extensions of physical theories may then better grasp the singularity of life.

Of course, this second term of the alternative is not necessarily associated with the particular and somewhat unilateral approach we have just been considering: other "axiomatic" points of view would doubtlessly enable us to use it. But the important issue we would like to stress, by means of something that is but an example of a possible approach, resides in the capacity of neglecting in principle neither the unifying perspective of mathematical formalization, nor the variability of the empirical phenomenality of that which makes the specificity and unity of life phenomena. It is possible to elaborate a conceptualization, which, without, however, falling into a physicalistic reductionism, does not renounce finding intrinsic formal relationships between the theory of physical (inert) materiality and the theory of biological (living) materiality. It goes without saying that, in accordance to what we have emphasized abundantly above, such a physicalistic reduction does not exist, at least for the time being. And this, not only for substantial reasons (even if matter – inert or living – is unique) related to the fact that the separation/bridging of the two disciplines involves the overstepping of a high number of organization levels and a redefinition of elementarity relatively to simplicity and to complexity, but also because there exists in fact many physical theories of biology – each of which can present very interesting illustrations for such or such a particular aspect – and because an eventual complete reduction would pose the question of knowing which would become the most relevant and predominant. Finally, one should also consider that, in spite of our effort towards a unified understanding by the geodesic principle, some key technical aspects of physics are still not unified: the field theories of (general) relativity and quantum mechanics, typically.

4.4 About the Possible Recategorizations of the Notions of Space and Time under the Current State of the Natural Sciences

In all the preceding, it has appeared more or less clearly that the concepts of space and time introduced and used in most contemporary theories had only remotely something to do with the spontaneous intuitions of these notions and with their actual employment within the classical frameworks of the natural sciences. Already, in order to account for this situation, we have been led to introduce what we have called "gnosiological complexes," referring to abstract structures that mobilize concepts such as those of the group or semi-group in mathematics or, in mathematical logic, of equivalence relations or of order relation. The most recent developments in physical theory, associated with more specifically biological considerations, would incite us to attempt to possibly go even further, or at least in a new direction, namely with regard to the notions of inside/outside and of the conceptual couple of internal/external.

From a biological point of view, we know that this couple is conceptually structuring since the introduction, by Bernard, of the notion of "*milieu intérieur*," which we have already evoked in Chapter 3. In what concerns space, that which pertains to an external environment, submitted to heteronomic laws that are physically and biologically relevant, and that which pertains to internal topologies (embryogenesis, development, anatomo-physiological organization), referring to autonomy, are thus dissociated.[8] These are dissociated at least since one of the issues is precisely to correctly describe their articulation.

In what concerns time, it is by no means less clear that two notions are confronted depending on whether we are dealing with processes of response to "external" stimuli (presenting characteristic relaxation times or periods, being of a purely physical nature, expressed in dimensional terms – seconds, hours ...) or, contrastingly, to internal regulation, of an iterative nature (heart beats, breathing, ...), represented in fact by these pure numbers mentioned earlier and which only acquire such a temporal dimensionality by the fact of the scalings to which they obey relative to the "size" (or "weight") parameter of organisms.

---

[8] In this vein, see the self-organizing approaches, which aim to account for this situation by introducing the concept of "organizational closure" (Varela, 1989), for instance, or the use which is made in the same perspective of modelizations by means of dynamical systems, as in Kaufmann (1993).

The new epistemological fact, which we have already evoked but which is now good to take again into consideration, is that physics, in turn, breaking with previous conceptions, is henceforth led to increasingly operate the same kind of distinction between that which is an issue of an "exteriority," the space and time of relativity, and that which pertains to a sort of interiority – fibers, of which the geometrical and symmetrical properties contribute to characterize the quantum behaviors and identities. This characteristic, already present in quantum mechanics as such, acquires considerable scope in quantum field theory and becomes truly determining in unification theories (between quantum physics and general relativity) which have recourse to string and superstring theories (introducing spaces of 10, even 26 dimensions, of which the exceeding dimensions are compactified[9]).

These approaches appear to be counter to usual intuition, but we very well know that mathematization confers to reference universes an increasingly counter-intuitive status. At the same time, it confers to "physical objects" themselves properties which seem increasingly strange, to the point of putting into question the usual concepts of causality and locality. Once more, we see the mathematical formalisms being developed (either diagrammatical or symbolic), while structural intuitions are elaborated, these being new intuitions without much relationship to previous ones, and which are both quasi inherent to the generativity of the mathematical principles of construction as well as being less and less empirical.[10] The problem which then arises is that of the rational articulation between these new types of intuitions and the factual results which present themselves in the usual universe of natural phenomena (of physics or of biology).

Conclusion In our brief inquiry concerning the mathematical principles currently appearing to be the most relevant in the attempt at rigorous characterization and at rational reconstitution within the natural sciences, it rather seems that both the theoretical results accumulated these last years as well as the introduction of new concepts related to renewed approaches led us to put into perspective the epistemological interpretation of

---

[9]In fact, and as we have already mentioned, the increase in dimensions of space-time as a reference universe in view of a unification between interactions is a method which precedes its employment within the framework of quantum theories (see Chapter 3).

[10]One will note that it is not the first time in scientific history that cognitive phenomena of this type occur. The passage from Ptolemy to Copernicus entailed a great loss of direct intuition: the sun so clearly revolved around the earth! But even the Galilean theory of movement and of dynamics clashed with the corresponding Aristotelian intuition and according to several inquiries, still continues somewhat to do so.

the status and of the determination of this objective knowledge. In contemporary physics, it is necessary to observe a radical rupture from previous analyses: besides quantum and relativity theories, issues related to critical phenomena, to non-linear dynamics and to complexity have opened a truly original field. The analyses of modes of construction of scientific objectivity find themselves comforted at the expense of radical empiricisms on the one hand, and of ontological realisms on the other. Manifestly, the practice of contemporary research as well as the presentation of its results and their conceptual interpretation articulates two moments which are increasingly indissociable: the constructive moment and the transcendental moment. The first is to be found in generalizing induction, the back and forth movement between theory, experience, and observation, the search for deducibility, the putting into form of hypotheses and conclusions. The second manifests itself as much in the intuitive movement which short-circuits the various stages (though this may mean having to restitute them later), by the passing to the limit which transforms facts into laws and laws into principles. The abductive reasoning which is associated with this process, the conceptual reversal which passes from the deductive to the prescriptive and which finds in these laws and principles the expression of their abstract conditions of possibility are part of this aspect.

Moreover and in this spirit, we wanted to show that it was possible to consider the possibility of a sort of mathematical reconstruction of these forms of intuition which are time and space, starting from the mathematical concepts of symmetries, of group, of equivalence on the one hand, of symmetry breaking, of semi-group and of order on the other hand. And if not accomplishing a complete reconstruction of these structures, we wanted in any case to open possibilities in this direction, using the results that the most recent physical science may produce in terms of invariance. However, let's reiterate that it is not a question of a complete and definitive objective reconstruction: in the same way that continuous mathematics, for instance, has and still undergoes many changes in its most variegated philosophico-mathematical determinations, from the sophistic paradoxes to the recourse to large cardinals. From Aristotle to Leibniz, Cantor, Dedekind, the continuous-discrete of the intuitionists (Bell, 1998) and of non-standardists (Harthong, 1983), we can expect the mathematical determinations of these fundamental structures which are space and time, and which are prerequisite conditions to all objective knowledge in the sciences of nature, to continue to develop in step with the theoretical advances in these disciplines such as the deepening of their mathematical foundations.

Earlier, when addressing superstring theory, we also mentioned, as examples, some spaces with high numbers of dimensions (10 or 26, of which most cannot be observed since they are compactified), and which are endowed with properties of supersymmetry (enabling the passage between fermionic and bosonic fields). We mentioned this in view of unifying quantum field theory with relativistic gravitational theory (see Bennequin, 1994).

It is, moreover, striking to observe in this regard that the issues raised by the theory of elementary particles are akin to those which proved to be the most relevant in terms of cosmology. Indeed, the states closest to the Big Bang are also those where the various interactions are physically unified and where the quantum constraints, most often associated with very short time or magnitude scales, relate to those of a gravitational nature, the latter being associated with the rather high densities on these very short scales. And this back to the origin, as the Big Bang itself is (hypothetically) attributed to a spontaneous symmetry breaking of the highly energetic quantum void. It is such results that have led certain authors (Weyl, 1918b; Cohen-Tannoudji *et al.*, 1986) to introduce the matter-space-time gnosiological complex, as mentioned earlier. This doubtlessly conduces to new refinements in the mathematical characterization of that which constitutes the physically relevant aspects of the study of material phenomena. It must be expected that progress in terms of a yet to be accomplished mathematization in the field of biology will lead to the introduction of a rigorous characterization of living phenomena, thus producing an even more unifying gnosiological complex and thus enabling us to reinterpret what significances the associated spatialities and temporalities could elicit.

It is, however, necessary, at this stage, to note that we are thus progressively engaged in a somewhat paradoxical situation from a conceptual point of view: on the one hand a transcendental type approach conduces us to consider space and time as primary data (forms of intuition, conditions of possibility of all sensible experience) and thus as strictly "empty" referential frameworks (pure forms). But on the other hand, and as with any construction of scientific objectivity, we are constantly led to reconstruct, according to theoretical advances, the mathematical structures which we will also call space or time. These are less and less empirical and increasingly determinant for the phenomena themselves, to a point where it is virtually impossible to distinguish between events and the "locus" of their occurrence. Each of these reciprocally determines the other, as is already spectacularly apparent in general relativity and in quantum field theory. Biology, as a science of processes, tends to increasingly foster such a representation. The distinction between forms of intuition and the mathematical intuitions which would come to "fill" them, doubtlessly conceivable within the framework of Newtonian physics and its absolutization of time and space, becomes less operative in the sense, at least, that the unifying mathematical constitutive determinations are increasingly so, even if this entails going against the first intuitions.

Chapter 5

Causes and Symmetries: The Continuum and the Discrete in Mathematical Modeling

Introduction

How do we make sense of physical phenomena? The answer is far from being univocal, particularly because the whole history of physics has set, at the center of the intelligibility of phenomena, changing notions of *cause*, from Aristotle's rich classification, to which we will return, to Galileo's (too strong?) simplification and the modern understanding in terms of "structural relationships" or the replacement of these notions by structural relationships. It is then an issue of the stability of the structures in question, of their invariants and symmetries (Weyl, 1927 and 1952; van Fraassen, 1989; see the previous chapter). To the point of the attempt to completely dispel the notion of cause, following a great and still open debate, in favor, for instance, of *probability correlations* (in quantum physics, see, for example, Anandan, 2002).

The situation is even more complex in biology, where the "reduction" to one or another of the current physico-mathematical theories is, as we have seen, far from being accomplished, if ever possible. Our point of view should henceforth be clear: the difficulties in doing this reside as much within the specificities of the causal regimes of physical theories – which, moreover, differ amongst themselves – as in the richness specific to the dynamics of living phenomena. Our approach has attempted to highlight certain aspects, such as the intertwining and coupling of levels of organization, which are strongly related to the phenomena of autopoiesis, of ago-antagonistic effects (in short, a dynamic opposition of "forces" or causes), of the hybrid causalities often mentioned in the theoretical reflections in biology (see Varela, 1989; Rosen, 1991; Stewart, 2002; Bernard-Weil, 2002; Bailly and Longo, 2003).

We will now return to some aspects of the construction of scientific objectivity, as explication of a theoretical web of relationships. And we will mostly speak of causal relationships, since causal links are fundamental structures of intelligibility. Our approach will again be centered upon symmetries and invariances, because they enable causes to manifest themselves, namely by the constraints they impose. In a strong sense, they thus present themselves as conditions of possibility for the construction of mathematical or physical objectivity.

Now, if mathematics is constitutive of physical objectivity and if it makes phenomena intelligible, its own "internal structure," that of the continuum, for example, as opposed to the discrete, contributes to physical and biological determination and structures their causal links. To put it in other words, mathematical structures are, on the one hand, the result of a *historical formation of meaning*, where history should be understood as the constitutive process from our phylogenetic history to the construction of intersubjectivity and of knowledge within our human communities. But, on the other hand, mathematics is also *constitutive of the meaning of the physical world*, since we make reality intelligible via mathematics. Particularly, it organizes regularities and correlates phenomena which, otherwise, would make no sense to us. The thesis outlined in Longo (2007) and which we further develop here, is that the mathematics of continua and discrete mathematics, the latter characteristic of computer modeling, propose different intelligibilities both for physical and living phenomena, particularly for that which concerns causal determinations and relationships as well as their associated symmetries/asymmetries.

In yhe final section, we will attempt to address the field of biology by questioning ourselves about the operational relevance and status of the concepts thus under consideration. But in this text, we will first propose to illustrate, in the case of physics, the situation which we have just summarily described, all the while returning to the reflection conducted in Chapter 4: this will enable us to "enframe" physical causality and to compare it to computational models and to biology.

5.1 Causal Structures and Symmetries, in Physics

The representation usually associated with physical causality is oriented (asymmetric): an originary cause generates a consecutive effect. Physical theory is supposed to be able to express and measure this relationship.

Thus, in the classic expression $\mathbf{F} = m\mathbf{a}$, we consider the force $\mathbf{F}$ to "cause" the acceleration $\mathbf{a}$ of the body of mass m and it would seem downright incongruous, despite the presence of the equality sign, to consider that acceleration, conversely, may be at the origin of a force relating to mass. Yet, since the advent of the theory of general relativity, this representation found itself to be questioned in favor of a much more balanced interactive representation (a "reticulated" representation, one may say): thus, the energy-momentum tensor doubtlessly "causes" the deformation of space, but, reciprocally, the curvature of a space may be considered as a field source. Finally, it is the whole of the manifesting network of interactions which is to be analyzed from the angle of geometry or from that, more physical, of the distribution of energy-momentum. An essential conceptual step has been made: to the expression of an isolated physical "law" (expressing the causality at hand) has been substituted a general principle of relativity (a principle of symmetry) and the latter re-establishes an effective equivalence (interactive determinations) where there appeared to be an order (from cause to effect).

Here is an organizing role of mathematical determination, a "set of rules" and a reading which is abstract, but rich in physical meaning. As we were saying in Chapter 4: causes become interactions and these interactions themselves constitute the fabric of the universe; deform this fabric and the interactions appear to be modified, intervene upon the interactions themselves and it is the fabric which will be modified.

We will first of all distinguish between *determinations* and *causes* as such. For instance, we will see the symmetries proposed within a theoretical framework as related to the determinations which enable causes to find expression and to act; in this they are more general than the causes and are logically situated as "prior" despite having been established, historically, "afterwards" (the analysis of the force of gravitation, as cause of an acceleration, preceded Newton's equation).

Let's then specify what we mean by "determination" in physics, enabling us to return to the causal relationships which we will examine extensively. For us, all these notions are the *result* of a construction of knowledge: by proposing a theory, we organize reality mathematically (formally) and thus constitute (determine) a phenomenal level as well as the objectivity and the very "object" of physics. We will therefore address first of all the "objective and formal determinations," particular to a theory.

More specifically, once given the theoretical framework, we may consider that:

D.1 The objective determinations are given by the *invariants* relative to the symmetries of the theory at hand.

D.2 The formal determinations correspond to the set of *rules and equations* relative to the system at hand.

To return to our example, when we represent the dynamic by means of Newton's equation, we have a formal determination based upon a representation of causal relationships, which we will call "efficient" (the force "causes" acceleration). However, when having recourse to Hamilton's equations we still have a formal determination, but one which refers to a different organization of principles (based on energy conservation, typically). It is still different to the optimality of the Lagrangian action, which refers to the minimality of an action associated with a trajectory. In this case, we have, for classical dynamics, three different mathematical characterizations of the events; and it is only with the advent of the notion of "gauge invariant" (that is, of "relativity principles") that these distinct formal determinations have been unified under an overreaching *objective determination*, related to the corresponding symmetries and invariants (manifested by transformation groups, such as the Galileo, Lorentz-Poincaré or Lie groups). A single objective determination then, for instance, the movement of a mobile with a certain mass, may account for (result from!) distinct formal determinations, based upon the concepts of force, of energy conservation and of geodesics, respectively. In the first case, the invariant is a property (mass), in the second, it is a state (energy), in the third it is question of the criticality of a geodesic (action, energy multiplied by time). If the final results of the mobile's dynamic may thus be the same, on the other hand the equations leading to them may take quite different forms unifying only under the even larger constraint of objective determinations (relating, in our example, to a mass in motion).

It is in fact the physical *objects* themselves which are the consequence of – given by – these determinations. More specifically, the physical objects are theoretically characterized by that which we designate, rather commonly, as properties and accessible states:

O.1 Properties (mass, charge, spin, other field sources, ...),

O.2 Accessible states, potential or actual (position, moments, quantum numbers, field intensity,...),

It is understood that their specific values essentially depend on empirical measurement. To highlight as simply as possible the difference we make between property and state, by means of their invariance characteristics, we may say that properties (which characterize an object) do not change when the states of the object change; conversely, if the properties change, it is the object itself which is modified.

These objective determinations thus constitute in a way the referential framework, at a given moment in time, to which are related *experience, observation and theory*, enabling us to interpret and to correlate them to each other. In themselves, and as we have just indicated, they thus do not completely characterize the objects they construct, but constrain – among other things by extricating invariants – properties and behaviors. Thus, for instance, they impose the fact that there is a mass (sensitive to the gravitational field), but without, nevertheless, fixing the magnitude of this mass or, as we shall see, the manifestation of fields such as the electro-magnetic field. We are therefore facing properties, which we may qualify as "categorical" and qualitative, but without necessarily specifying the associated quantities which quantitatively characterize the object in direct relationship to the measurement. This is also the case for that which we call accessible states: their structure is qualitatively characterized, but the fact that the system quantitatively attains one or other of these theoretically determined possible states depends on empirical factors.

Why distinguish here between properties and accessible states? It would enable us to understand as *cause*, in the traditional sense (which after Aristotle we will call "efficient cause"), all which affects (can modify) states; while we may consider that in the traditional approach, *the invariants of efficient causal reduction* are constituted by the set of properties. However, these very properties participate to a causality, which we shall relate to "material" causality.

So let's attempt to refine the analysis, not only by distinguishing between different types of "causes" but also by trying to affect the distinct elements of objectivity. Let's agree that, relative to the effect of an object upon another:

C.1 The **material cause** is associated with the set of *properties*;
C.2 The **efficient cause** is correlated to the variation of one or more *states*.

We can recognize here a revitalization of Aristotle's classification, so dear to Thom. In fact, if we want to maintain a parallel with Aristotelian categorization, let's observe that we have called formal determination what the modern interpretation of the philosopher would designate as "formal cause" (that is, that which corresponds to the set of theoretical constraints which define and measure the effects of other causes – laws, rules, theories, ...)[1]. In our approach, it is the determinations, formal and objective, which produce the specification of objects, by means of the notions of properties and states (of which the structures and variations participate in the material and efficient causes, respectively). With regard to the causes, we will preserve the Aristotelian terminology, although material causes may be classified as "material structures." Indeed, a change in properties changes an object, as we mentioned earlier, but, at the same time, it induces – it causes! – a change of states. For example, a change in mass or charge, *in an equation*, modifies the values of the acceleration or of the electrical field.

5.1.1 *Symmetries as starting point for intelligibility*

From the point of view we have just developed, may we consider that constraints of symmetry stem from causal constraints? According to our distinction and as we have just specified, symmetries emerge from the determinations (under the form of systems of equations, typically) where the causes manifest. Their greater generality thus also imposes itself through the relation to laws corresponding to the formal determinations (which, for example, take such or such expression according to the selected gauges). To put it lapidarily, using the example we will discuss below (Intermezzo): the phase's global gauge invariance *determines* the charge (a property) as a conserved quantity of the theory and its local invariance *determines* the existence of the electromagnetic field (a state) under the form of Maxwell's equations. The interactions, described by these equations, may, but only afterwards, be considered as giving us the *causes* (material or efficient) of the observed effects.

---

[1]In the debate with Prigogine concerning determinism, Thom highlights the role of structural stability, even within the framework of highly unstable dynamics (the forms are maintained, all the while being deformed). It is the equations of the dynamic which determine their possible evolution (as formal causes – determinations, for us). On the other hand, Prigogine highlights the play between locally stable structures and global systems where small, amplified fluctuations induce the choice of one of these evolutions. While preserving his new view on Aristotle's finesse, but in a different way than Thom, we do not attribute to these different notions of causality an ontological hierarchy of the Platonic type, where the formal determinations (causes) *ontologically* precede the other causes.

In fact, and since Galileo, what we usually characterize as "causes" seem to correspond mainly to efficient causes, whereas, as we have just seen, the "determinations" seem to rather present themselves as a source common to causes which would derive from them (including material and formal). This results in that we may consider, "transcendentally speaking," the determinations, the symmetries, namely, to present themselves as conditions of possibility for the causes to manifest.

Now it appears to us that the natural sciences, with the exception of the biosciences, may be part of the conceptual framework we have just drawn out, including that which corresponds to extremalisation rules (the geodesics of the Lagrangian), which appear, but wrongly so in our opinion, to confer a tinge of *finality* to the processes which they model. As a matter of fact, a geodesic is not an "optimal path towards" (the light doesn't "aim at a target" by the shortest path). The path is optimal by integration of "local conservation properties" (the energy variation is optimized at each instant). In short and to make it simple, a physical body moving along a line preserves the *tangent constant*, point by point, without looking towards a target. This is Galileo inertia, the first understanding and instance of a geodesic and of energy conservation principles. It is only with living phenomena that the taking into account of a sort of "final causality" (to put it once more in Aristotelian terms, see Rosen (1991); Stewart (2002)), that we have characterized elsewhere as a "contingent finality" (see also section 5.3 below) and as locus of "meaning" for any living phenomenon, really becomes relevant. It is this which we will attempt to examine later, in paragraph 3.

5.1.2 *Time and causality in physics*

We have thus attempted to specify, very generally, the notions of objective determination, of object and physical cause, from the notion of symmetry and, more specifically, from the notion of invariance with regard to the given symmetries.

Let's also observe that, for about a century, in physics, the laws of conservation, as formal determinations, have been understood in terms of spatio-temporal symmetries; for instance, the conservation of angular momentum is correlative to the symmetry of rotation (it is Noether's theory which is at the origin of this great theoretical and conceptual turning-point, see Chapter 4 and the Intermezzo below).

But at this stage and before continuing, it would seem appropriate to introduce a distinction in the view of clearing some possible confusion with regard to the representation of causality and to the reasoning that one may entertain about it. We propose to distinguish,[2] namely in the case of efficient causality, between *objective causality* and *epistemic causality*.

Objective causality is associated, in our opinion, with a rather essential constraint, which is constitutive of physical phenomena, and which is the irreversible characteristic of the unfolding of time (what we call the "arrow of time"). But even in the case where temporality does not explicitly appear, it continues to underlie any change, any process as such – including that of measurement – and constitutes in this respect a foundation to any conceptualization, observation, or experience, from the moment that such a process is considered. That is to say, this time from the angle of causal analysis, that *time is constitutive of physical objectivity*.

In contrast, epistemic causality is to be considered as independent of an arrow of time. For instance, the analysis of a phase transition as a function of the value of a parameter (such as temperature) does not refer to any specific temporality. It is in a way the atemporal and abstract variation of the parameter that "causes" the transition, be it in one direction (for instance, from liquid to solid) or the other (from solid to liquid). At this level, the invoked structure of causality (effect of the change of temperature on the state of the system) remains independent of the time factor, even though, at another level, it is indeed over time that the effective variation of this parameter occurs – in one direction or the other. This is also the case in the very simple example which is the law of perfect gases ($pV = RT$, where p represents pressure, V the volume, T the temperature, and R Joule's constant). This law is independent of time and one may conceive of various "causes" as the origin of a variation in volume, for instance, leading, under constant temperature, to a variation in the pressure associated with the occurrence of a chemical reaction. The concomitant (and symmetrical, given the equational relationship) variations in volume and pressure may be considered as the causes – of the epistemic type – of one another (in fact, these variations may be said to be "correlated"), in contrast to an objective causality – temporalized, this time – which would find its source in the temporal unfolding of this chemical reaction at the origin of the considered variation in volume (our distinction may possibly help in understanding the discussion in Viennot (2003; appendix)).

---

[2] As we have already done on the occasion of the approach and the deepening of the concept of "complexity" (Bailly and Longo, 2003).

May such a distinction between the objective and the epistemic, which seems to clearly correspond to a reality in the case of an efficient causality (associated, let's remember, with modifications in the states of a system, as we have just illustrated) still be applicable to material causality?

It does appear that for material causality, one may find examples demonstrating that such is the case, inasmuch as the properties in question have different expressions depending whether they are related to their own specific system or to an external reference. This is the case in relativity, for example, where the mass (or life-span) of particles depend on their speed with regard to the laboratory reference: in the *internal* system, the rest mass remains a characteristic property of the particle's very identity (m_0), whereas within a referential animated by a speed v with regard to the system as such, the mass takes on an epistemic character, of which the measurement is $m = m_0/(1 - v^2/c^2)^{1/2}$, where c represents the speed of light (for light itself, this is also what enables us to consider that the photon's mass is null, while its energy is non-null and Einstein's relation establishes a direct relationship between mass and energy). Likewise, the "efficient mass," which we calculate following the process of renormalization (which, in order to eliminate infinities from the calculi of perturbation, integrates with the mass some classes of interaction), takes an epistemic character with regard to the mass itself which preserves its own objective character. In this sense, we may consider that the properties, which are located at the source of material causality, retain an objective character in their internal system all the while acquiring an epistemic character if we relate them to different referentials.

Because we take the arrow of time into consideration while characterizing efficient objective causality, we distinguish certain trends in relativistic and quantum physics that exclude such an arrow, in order to preserve any relationship by symmetry. In these approaches, the causal relationships are replaced by other concepts, for instance, in quantum mechanics, by probability correlations (see Anandan, 2002, among others). The reason for this differentiation, beyond the elements of analysis we have just exposed, appears to be crucial in an epistemological respect: we will indeed often refer to dynamic systems (thermodynamic and of the critical type) and will also address certain aspects of biology. Now, there is no analysis of these systems, even less of living phenomena, which can be performed without taking into account the existence of an arrow of time. Particularly, there would be no phylogenesis, no ontogenesis, no death, ... in short; there would be no life without time, oriented and irreversible. The processes of life impose an arrow of time, be it only for the thermodynamic effects to which they participate; but it even appears unavoidable to go further, because these processes require a new way of looking at complex forms of temporality, of clocks of life with causal retroactions due to intentional aims, to expectancies and previsions, characteristic of perception and action (see Chapter 3).

To conclude, our mathematical point of view is that objective determinations are given by symmetries and an efficient cause *breaks* some of the latter, be it, from an objective viewpoint, only that symmetry which is associated with the arrow of time. Reciprocally, irreversible phenomena (bifurcations, phase changes, ...), which are therefore oriented in time, may be read as symmetry breaking correlated to (new) causal relationships. Symmetries and their breaking therefore remain the starting point for any theoretical intelligibility.

More specifically, we will attempt to understand some causal relationships as symmetry breaking, in a very general and abstract sense. This will, among other things, enable us to lay the basis of a coherent foundational framework for the analysis of the different causal regimes proposed by continuous mathematics in comparison to those of discrete arithmetic. This will therefore consist of a mathematical view upon the constitutive role of mathematics in the construction of scientific objectivity; through this approach, we aim to grasp the importance of our digital machines in this construction, since these machines are the practical realization of the arithmetization of knowledge.

The final reflection regarding biology will bring us back to natural phenomena, in all their causal specificity. Of course, computerized modeling, in biology as in physics, remains a fundamental issue. It is precisely for this reason that it must be based upon a fine analysis of the different structures of relationships, particularly causal relationships, proposed within the various theoretical frameworks (physical, biological, of discrete mathematics).

5.1.3 *Symmetry breaking and fabrics of interaction*

It is thus by the means of mathematics that we organize causal links; mathematics makes intelligible and unifies, particularly via symmetries, certain phenomenal regularities, at least those of classical physics, both dynamical and relativistic systems. But mathematics also makes explicit the symmetries in relation to which probability correlations are quantum invariants.

In the dynamic and relativistic cases, the geodesic principles governing the evolution of systems apply to abstract spaces, "manifolds" endowed with a metric where, as we were saying in Chapter 4, *symmetry transformations*[3] leave equations of movement invariant[4] (typically, the trajectories defined by Euler–Lagrange equations, see Chapter 4). It is in this sense that these theories base themselves on invariants with regard to spatio-temporal symmetries: if we understand the "laws" of a theory to be "the expression of a geodesic principle within a suitable space," in the sense of Chapter 4, it is these abstract geodesics which are not modified by transformations of symmetries.

Let's return to the most classical of physical laws: the equation $\mathbf{F} = m\mathbf{a}$ is symmetrical, as an equation. As we observed above, it is its *asymmetrical* reading which we associate with a causal relationship: the force $\mathbf{F}$ *causes* the acceleration $\mathbf{a}$ (the equation is read, so to speak, from left to right). We thus break, conceptually, a formal symmetry, equality, in order to better understand, following Newton, a trajectory (and its cause). More specifically, the equation *formally determines* a trajectory of which $\mathbf{F}$ appears as the *efficient* cause (it modifies a state, all the while leaving invariant the Newtonian mass, a property). In this sense it becomes legitimate to consider that the equation contributes to the constitution of an objectivity (the trajectory of the mobile), whereas its oriented reading (and the efficient causality it thus expresses) constitutes an interpretation and refers to an epistemic regime of causality.

We therefore propose to consider that each time a physical phenomenon is presented by an (a system of) equation(s), *a breaking in the formal symmetry* (that of equality, by an oriented reading) *makes explicit an epistemic regime of causality*. Particularly, the symmetry breaking in question may be correlated to an efficient cause which intervenes within the formal framework determined by the equation.

Of course, this breaking is not necessarily unique (that by which, namely, it manifests its epistemic character). For example, as we have just evoked in the preceding paragraph, one can read causally and from an epistemic standpoint $pV = RT$ from left to right and vice versa. By reference to relativistic systems, we have already read the equation $\mathbf{F} = m\mathbf{a}$ (or more exactly, its relativistic equivalent), inversely, while highlighting the fact that, reciprocally, the curvature of a space may be considered as a field source. This interpretive reversal, which reorganizes phenomena

---

[3]Objective determinations, in our language.
[4]Formal determinations, for us.

radically, is legitimate; indeed, in our spatial manifolds, *the transformations* (of gauges), which enable us to pass from one referential to another, are supposed to leave invariant the equations of movement, and by doing so they preserve the *symmetries*, but without necessarily preserving the asymmetrical readings of the formal determinations (among which is the epistemic causal reading we have just discussed).

We thus propose to consider formal interactions, organized by the asymmetrical structures of equations, but also (efficient) causes, which can be associated with possible asymmetries in the reading of these very equations. Let's observe, once more, that certain coefficients, such as the mass m in $\mathbf{F} = m\mathbf{a}$, are correlated to that which we have categorized as material causes (whereas acceleration is correlated to states). And, when an "external" cause (efficient or material) is added to a given determination (equations of evolution), the geodesics of the relevant space are deformed and symmetries associated with this space may be broken, following the variation of states and properties.

Now, this mathematical intelligibility, conceptual fabric of symmetries and asymmetries which correlates the regularities of the world, is constitutive of physical phenomena as well as of scientific objectivity. As we shall see, they profoundly change if the world's proposed reading grid is rooted in continuous or in discrete mathematics. And they must subsequently be enriched, if one hopes to better conceptualize certain phenomena pertaining to life.

Intermezzo: Remarks and technical commentaries

Inter. 1: More on symmetries and symmetry breaking in contemporary physics

Let's consider the previously analyzed three great types of physical theories, which are the relativistic, the quantum and the critical types (dynamic and thermodynamic systems).

Relativistic theories are essentially tributary of *external* symmetries (sets operating over space-time). Classical mechanics already presents these relativistic traits, with its constraints of invariance under the Galileo group (within the Euclidean space), but it is especially with classical electromagnetism and special relativity that symmetries begin to play a determining role under the Lorentz–Poincaré group (group of the rotations and translations within a Minkowski space). Regarding general relativity and cosmology, it is the group of the set of diffeomorphisms of space-time which plays the determining role. The corresponding symmetry breaking principally manifest itself through phenomena of dissipation, or of the arrow of time.

Quantum type theories for their part mobilize essentially *internal* symmetries operating on the fibers of the corresponding geometric "fibrates": it is the gauge sets which generate the gauge invariances and which present themselves as Lie groups (continuous groups). In quantum field theory, the most important symmetry breakings (Goldstone, Higgs fields) are considered as sources of the masses of quanta.

Critical type theories constitute theories par excellence of symmetry changes (namely by breakings): it is phase transitions, spontaneous symmetry breaking (or, conversely, of the apparition of new symmetries), of which the effects are processed this time by means of the renormalization semi-group process in order to characterize the critical exponents and to established rules of universality which constitute, in a certain way, over the classes of equivalency which they bring forth, the basis of new relativities and symmetries (very different symmetries may present critical exponents, and thus behaviors, which are identical, depending only on parameters as general as the magnitudes of the prolongation spaces or that of the order parameters).

We will also mention the fact that the unification processes, in the relativistic theories as well as in the quantum theories, or even between themselves, involve the expansion of the concerned symmetry groups (often at the same time as the spaces within which they operate). In other words, the current way for correlating different structures of determination in physics is to refer to new theories that bypass the specific ones, in particular by more symmetries (and associated groups) and symmetry breaking.

Inter. 2: From Noether's theorem and physical laws of conservation

One of the principal foundations of the role of symmetries for physics can be found in Noether's theorem (see also Chapter 4), according to which any transformation in symmetry, operating upon a Lagrangian and conserving the equations of movements, associates conserved quantities. By a more precise analysis, one may observe that this theorem narrowly couples such laws of conservation – physical invariants, that is, objective determinations – to indeterminations of reference systems (space-time, fibers) by the fact of the principles of relativity and of the symmetries supposed to operate there (for instance, the impossibility of defining a temporal or positional origin, an origin of phases, etc.).

One of the simplest and most spectacular cases we may evoke in this regard is that of quantum electrodynamics, for which the gauge group is the phase group $U(1)$. In this case, it is required that the form of the density of the Lagrangian remains invariant under the multiplication of the state vector by a phase term $(\exp(iL))$. The global gauge invariance (L *independent* of the position), conduces, by the application of Noether's theorem, to the conservation of a quantity which we identify as the charge and which corresponds, according to the classification which we propose, to a "property," that is, to a material characteristic. Moreover, the local gauge invariance (L *dependent* of the position) requires, in order to re-establish the broken Lagrangian covariance, the introduction of a gauge potential, where there results a gauge field which is no other than the electromagnetic field itself, expressed by the Maxwell equations (the gauge potential corresponding to its vector potential), which corresponds itself to the source of an efficient causality. Thus, it is indeed the indetermination of any phase origin (an aspect of the referential universe) which very strongly determines the conservation of the charge (an aspect of the determination of the physical object) and, most of all, for local invariance, determines the electromagnetic field itself, nevertheless, generally interpreted as a "cause" of electromagnetic phenomena. Here, one may also notice (but we will not proceed further in the analysis, at this stage) that it is the global gauge invariance which finds itself to be coupled with a property (the charge), whereas the local invariance is coupled with a phenomenon intervening upon the states (with an effect of efficient causality): the field. So there are two forms of invariance, with regard to spatiotemporal symmetries, which we understand as objective determinations of a property and of a state, respectively.

In fact, in theoretical work as in the quest for unification, it is indeed these properties of symmetry – these forms of indetermination of the referential universes – which play an essential heuristic role in the determination of physical phenomenality. As if we were passing from the prevalence of the representation by *efficient causality* to that of a representation by *formal determination* with its symmetries and equational invariants. It is appropriate to recall here the remark, already quoted, by Chevalley in his preface to Van Fraassen's book, where it is a question of "*substituting to the concept of law, that of symmetry.*"

We will also note, besides the continuous symmetries that we have mainly evoked, the important role played by discrete symmetries, as in the CPT theorem, according to which the result of the three transformations T (reversal of time), C (charge conjugation: passage from matter to anti-matter), P (parity, mirror symmetry in space) is conserved in all the interactions, whereas we know that P is broken by chirality in the weak interaction, and that CP is broken in certain cases of disintegration (which led Sakharov to see in this a reason for the weak prevalence of matter over anti-matter and therefore the existence of our universe). We may understand here the particularity of the approaches, in quantum physics, which do without the arrow of time (Anandan, 2002, for instance). The breaking of the CP symmetry is not taken into consideration, enabling us to have no asymmetry in the T transformations, and therefore to not have an oriented time factor.

With regard to spontaneous symmetry breaking, we have also evoked phase transitions and, from the quantum viewpoint, the Goldstone fields (for the global level) and the Higgs fields (for the local level), supposed to confer to particles their masses. But it is necessary to emphasize that from a cosmological standpoint, the decoupling of the fundamental interactions from each other (gravitational, weak, strong and electromagnetic), also constitute such breaking: they then correspond to differentiations, which have enabled our material universe to evolve into its current form. Without mentioning the fact that the Big Bang itself may be considered as the very first symmetry breaking (due to quantum fluctuations) of a highly energetic void.

But it is doubtlessly with regard to living phenomena that this symmetry breaking plays an eminently sensitive role. Thus, Pasteur, let's recall, who had lengthily worked on the chirality of tartrates, did not hesitate to assert: *"Life as it presents itself to us is a function of the asymmetry of the universe and a consequence of this fact."* More recently, dynamic models involving sequences of bifurcations have also been proposed to represent processes of organization of which living phenomena could be the locus (Nicolis, 1986; Nicolis, Prigogine, 1989).

5.2 From the Continuum to the Discrete

Differential and integral equations, as limits, but also variations and continuous deformations, are present everywhere, in the physico-mathematical analyses that we have evoked. From Leibniz and Newton to Riemann, the phenomenal continuum, with its infinity and its limits in action, is at the center of mathematical construction, from infinitesimal calculus to differential geometry: it constitutes the space of meaning for the equations (formal determinations) of which we have spoken, the structure underlying any spatial manifold (Riemannian). However, a discretization, or a representation that is finite, approximated, but "effective," should be possible. It is the dream, implicit in Laplace's conjecture, which will find its continuation in the foundational philosophy of arithmetizing formalisms. If, as Laplace had hoped for the solar system (see Chapter 3), to a small perturbation always responds a consequence of the same order of magnitude (except in critical situations, cases which are "isolated" – topologically – such as a mountain peak, of which Laplace was well aware), then today it would be possible to organize the world by means of well-delimited little cubes (corresponding to the approximation of digital rounding; to the pixels on our machines' screen) and to proceed to the arithmetic calculi upon these discrete values (the encoding of pixels by integers, sequences of 0s and 1s), which would then provide a "complete" theory (any statement concerning the future and the past would be decidable, *modulo* the concerned approximation). Indeed, arithmetical rounding, which associates a single number with all the values contained within a "little cube," does not perturb the simulation of a linear or Laplacian system, because the approximation which is inherent to it is preserved (modulo a linear growth) over the course of the calculi, just as over the course of physical evolution. Let's explain ourselves, because a whole philosophy of mathematical foundations and, in fact, of nature, stems from this approach, with its own view on causality and determination.

5.2.1 *Computer science and the philosophy of arithmetic*

Digital computers are changing our world, by means of the very powerful tools for knowledge they provide us with and by the image of the world they reflect. They participate in the construction of all scientific knowledge via simulation and the elaboration of data. But they are not neutral: their theory, as formal machines, dates back to the 1930s, when effective computability, a theory of functions upon integers of integral value, imposed itself as the paradigm for logico-formal deduction. Induction and recursion, arithmetic principles, are at its center. Our arithmetic machines and their techniques for the digital encoding of language (Gödelization) thus derive from a strong vision of mathematics, of knowledge in fact, rooted upon arithmetic; the latter has been proposed as a locus of certitude, and of the absolute (the integer number, "an absolute concept," for Frege), as a locus of the possible encoding of any form of knowledge ("of all which is thinkable," Frege), of geometry in particular (Hilbert, 1899), as an organizing theory of time and space. And certainty would be attained without preoccupation for the revolution caused by the geometrization of physics within non-Euclidean continua, with their variable curvatures, those of Riemann geometry (a "delirium," with regard to intuitive meaning – Frege *dixit*, 1884, see Chapter 1); without this incertitude of determinism deprived of predictability, characteristic of the geometry of dynamic systems since Poincaré. So a philosophy of arithmetic imposed itself upon foundational reflection, all the while departing from the new physics, which will mark the XXth century. And it proposes that we read the world *modulo* an arithmetic encoding, the same one enabling us to construct, from the world, the basis for modern digital data.

For this reason, the analysis of the constitution of intelligibility and of meaning, as an intrication of mathematics with the world, is not traditionally part of foundational analysis in mathematics. The mathematical logic of Frege and Hilbert, with the profoundness of its achievements and the force of its philosophy, has led us to believe that any foundational analysis could be reconducted as the analysis of an adequate logico-formal system, a logical system (Frege), or a finite collection of sequences of meaningless signs (Hilbertian school), of which the meta-mathematical investigation would then become an arithmetic game (following the digital encoding of any finitary formal system), perfectly removed from the world. And, since Hilbert, as we have seen, the *formal coherence* of these calculi of signs claims to provide the sole justification of these systems, even those of the geometry of physical space and of theories of continua, as they become reduced to arithmetic.

This has definitely separated mathematical foundations from the foundations of other sciences, including physics, despite the roles of construction and reciprocal specification between these two disciplines, having a common constitution of meaning. With regard to biology, the foundational interaction has been lesser, for the time being, following the lesser mathematization of this discipline. However, the ideology of the construction of the computational model as the main explicative objective has already marked the interface between mathematics and biology, all the while forgetting the strong commitment to structuring the world, implicit in its computational arithmetization; and few discussions have attempted to correlate the foundations of the arithmetizing theories to those of the theories of life (see Longo and Tendero, 2007). This epistemological separation makes difficult the interdisciplinarity and applications from one discipline to another, because foundational dialog is a condition of possibility for a thought-out interdisciplinarity, a starting point for a parallel constitution of concepts and practices and for a common formation of meaning.

5.2.2 *Laplace, digital rounding, and iteration*

So let's return to the "bifurcation" that happened: on one side, the arithmetization of the foundations of mathematics (from Frege and Hilbert, although in different frameworks), and on the other, the geometrization of physics (Riemann and Poincaré, in particular). The two branches have been quite productive: on the one hand we have the theory of effective computability and, therefore, our extraordinary arithmetic machines and, on the other hand, two fundamental aspects of modern physics. The first branch of the bifurcation, however, in its foundational autonomy, has continued to base itself upon Newtonian absolutes (Frege) and upon the Laplacian determination (Hilbert), the one which involves the predictability (and which has its counterpart, in meta-mathematics, in the "*non ignorabimus*," Hilbert's decidability: once a mathematical statement is well formalized, one must be able to demonstrate either it or its negation – to falsify it).

It is actually quite clear that Laplace's hypothesis explicitly and first of all aims for predictability ("any deterministic system is predictable"; that is, in a formally determined system, any statement – concerning the future/past – is decidable). However, it bases itself precisely upon this "conservational" interpretation of perturbation evoked earlier: Laplace is very well aware that physical measurement always constitutes an interval (it is necessarily approximated), but he believes that the solutions for the world's systems of equations, approximated if necessary by means of series (of Fourier), will be "stable" with regard to small perturbations, particularly those of which the amplitude remains below possible measurement. The perturbation, which by "almost insensible variations," could even induce quite important secular changes – in his words, should not impede the stability of the solar system (see Chapter 3). It is this which guarantees predictability: in a system which is deterministic (therefore, in principle, formally determined by means of equations), predictability is ensured by the resolvability of the system and/or the *preservation of the approxima-*

tions under certain conditions (given the values of the initial conditions, with a given approximation, it will be possible to describe the system's evolution by an approximation of the same order of magnitude). There is the conceptual (and historical) continuity, which we have already addressed, between Laplace's conjecture and the myth of the arithmetization of the world: approximation and rounding (discretization) do not modify the evolution under consideration, physical and simulated.

Yet it is nowhere like this. Even within a system, which is (relatively simple and) explicitly determined by equations (the symmetric structure of formal determinations: the nine equations of three bodies within their gravitational fields, for example), unpredictability arises, Poincaré has explained. What happens? Even within such a simple system, almost everywhere small perturbations may give rise to huge consequences; in fact, "small dividers" (which tend towards 0) within the coefficients of approximating series (of Lindstedt–Fourier) amplify the slightest of variations in the initial values. Ninety years later, this phenomenon will be defined as "sensitivity to the initial conditions" (or at the border, or "at the limits"). Particularly, even perturbations of which the amplitude is below the threshold of possible physical measurement may, after a certain amount of time, produce measurable changes.

So, in our interpretation, a perturbation, a "small force" which perturbs a trajectory even below that which is measurable, *breaks an aspect of the symmetry* described by the equation of the system's evolution; it is the cause (efficient or material) of a variation in the initial conditions, which may produce observable consequences, even very important ones. Sometimes, it may even be question of a fluctuation, such as a local or momentary breaking of the symmetry internal to the system, without the influence of "external" causes: in the Intermezzo, we have evoked the Big Bang in cosmology, as the very first symmetry breaking (due to quantum fluctuations) of a highly energetic void. Once again, it is a broken symmetry that is the *material or efficient cause* of a specific observable evolution, that of our universe.[5]

---

[5] According to the Curie principle, "the symmetries of the causes can be found in the symmetries of the effects." In the approach followed here and as we have observed, one would say that in these cases, at the level of the observable, this is not the case: to an apparently symmetrical initial situation may follow an observable evolution which does not reproduce the same symmetries, following a breaking of symmetry, initial or at the edges, of which the amplitude, initially, is below the threshold of possible measurement (therefore non-observable). In these cases, therefore, certain symmetries are not conserved when passing from (observable) causes to (observable) consequences.

Now, the intelligibility of these phenomena, present at the center of modern physics, is conceptually lost if we organize the world by means of the exact values that arithmetical discretization imposes. Or, rather, and here lies our thesis, we obtain a different intelligibility. Particularly, the perturbation or fluctuation, which have their origin in efficient or material causes, and which manifest *below the proposed discrete approximation*, elude arithmetic intelligibility, or are neglected in favor of a forced stability of phenomena. And arithmetic calculus shows us the passing from one state to another by little iterated jumps of trajectories that are imperturbable, because perfectly iterable. Better, it shows us trajectories which are affected by their own intrinsic perturbation, at each increment of calculus, and always identically iterated and iterable: *the rounding-off*. A new cause, our computational invention, which, projected into the world, becomes a relevant cause (efficient or material) with regard to the properties and states of a system. Because digital rounding modifies the simulated geodesics and may even change, in certain cases and in its own way, the conservation phenomena (of energy, of moment, ...) by breaking the associated symmetries. We will return to this point.

In what sense then, do we obtain, when we superimpose upon the world an arithmetic grid, a "forced stability," as well as evolutions and perturbations which are very specific and "iterable" at will? We will understand this thanks to the digital computer, because, when this arithmetic machine is used as a *model of the world*, it organizes the world according to its own causal regime, its own symmetries and symmetry breaking. In fact, the digital simulation of a physical process is constitutive of a new objectivity, to be analyzed closely because very important, due to the mathematical tools which are at its center: the arithmetic calculi and discrete topology of its digital databases and of its working memory space, exact and absolute.

It is clear that our analysis does not aim to oppose what would be an "ontology" of continua to what would be an ontology of discrete mathematics (we are not defending the idea according to which the world itself would be continuous as such!). We are rather attempting to highlight the difference of the views proposed by discrete mathematics in relation to those proposed by continuous mathematics, in our efforts to make the world intelligible. It is the constructed objectivity of mathematics which changes and not, we repeat, any ontology.

Moreover, the relative incompleteness of computational simulation, which we emphasize here, goes hand in hand with the mathematical incompleteness of arithmetic formalisms which, also, is relative (to the practice of mathematical proof, in this case, see Chapter 2). But incompleteness does in no way mean "uselessness": to the contrary, we emphasize the need for a fine conceptual analysis of algorithmic methods, precisely for the essential and strong role they play today in any scientific construction.

5.2.3 Iteration and prediction

Computers iterate; that is their strength. From primitive recursion, at the center of the mathematics of computability, to the software application, the program, the sub-program, re-run a thousand times, a billion times, once every nanosecond, all reiterate with absolute exactitude. For this reason, there is no randomness as such in a digital world: (pseudo-)randomness generators are small programs, perfectly iterable, which generate periodic sequences endowed with very long periods (they are functions iterated upon finite domains).

In the algorithmic theory of information, we call random any sequence of integers for which *we do not know* a generating program shorter than the sequence itself. That is, that we do *not see* any sufficient regularities within the sequence to be able to deduce a rule by which to generate it. This definition identifies with randomness the informational characteristics of a series of throws of dice or of roulette, in fact, their incompressibility. This identification, applied to algorithms, (to pseudo-)random generators for instance, leads to a confusion between an epistemic notion of randomness, specific to physics, and "randomness by incompetence," (the programmer has not told us how the program is designed, usually a one-line program). And iteration reveals the trick: if we re-launch a programmed (pseudo-)randomness generator, using the same initial values, we will obtain the same sequence, exactly. On the other hand, dynamical systems give us the good (epistemic, see the chapter 7) notion of randomness: a process is random if, when we iterate it with "the same" initial conditions, it does not follow, generally, the same evolution (dice, roulette, planetary systems having at least three bodies, if we wait long enough). The whole difference lies in the topological signification of this notion of "same" (*same* discrete values and *same* initial conditions): a digital database is discrete and *exact*; whereas physical measurement is necessarily an interval.

In short, in the mathematical universe of effective computability, there is no randomness as such, at best there is incomprehensible information (which may provide a good "imitation," see below, of randomness). And one may say synthetically with regard to our approach: the time of calculation processes is subject to a "symmetry" in terms of iterability (identical repeatability), which does not have an absolutely rigorous meaning in the physical world and even less so in that of living phenomena. This iterability is essential to computer science: it is at the center of software portability, therefore, of the very idea one may have concerning software; that one may transfer it onto any adequate machine and run it and re-run it identically as often as one wants. And it works, in fact.

Of course, computers are in the world. If we come out of the discrete arithmetic internal to the machine, we may embed them in physical randomness (epistemic – dynamic systems – or intrinsic – quantum physics, see chapter 7). We may, for instance, use temporal shifts within a network (a distributed and concurrent system, see Aceto *et al.* (2003)) upon which humans also intervene, randomly; or using little boxes, sold in Geneva, which produce 0s and 1s following quantum "spin-ups/spin-downs." But normally, if you run the simulation of the most complex of chaotic systems, a Lorentz attractor, a quadruple pendulum and iterate with *the same initial digital data*, you will obtain the same phase portrait, the same trajectory. The same initial data, there is the problem. As we have emphasized earlier, this physical notion is conceived *modulo* possible measurement, which is always approximated, and the dynamic may be such that a variation, including below the threshold of measurement – the material or efficient cause – (almost) always generates a different evolution. On the other hand, in a discrete state machine, "the same initial data" signifies "exactly the same integers." This is what leads Turing to say that his logico-arithmetic machine is a Laplacian machine (see Turing, 1950; Longo, 2002). Like Laplace's God, the digital computer, its operating system, has a complete mastery over the rules (implemented in its programs) and a perfect knowledge of (access to) its discrete universe, point by point. As for Laplace's God, "prediction is possible" (Turing, 1950).

And so thus is the philosophy of nature implicit in any approach, which confounds digital simulation with mathematical modeling, or which superimposes and identifies algorithms to the world. Discrete simulation is rather an *imitation*, if we recall the distinction, implicit in Turing, between model and imitation (see Longo, 2002). Very briefly: a *physico-mathematical* model tries to propose, by means of mathematics, the *constitutive* formal determinations of the considered phenomenon; a functional imitation only produces a similar behavior, based, generally, upon a different causal structure. In the case of continuous *vs* digital modeling, the comparison between different causal regimes is at the center of this distinction.

Turing has the huge merit of having invented the machines and of having, in 1950, when he abandoned the myth of the great digital brain, highlighted the difference between *computational imitation*, the game which should demonstrate his machine to be indistinguishable from a woman (modulo the sole intermediary of a written interface, (Turing, 1950)), and *modeling*. In fact, in his 1952 article on morphogenesis, he presents a *model* of chemical actions-reactions which generates forms and which bases itself upon a system of differential equations: tiny variations generate the variety of forms in certain natural phenomena (Turing calls this sensitivity "exponential drift," a quite relevant and original name! The notion of "sensitivity to the initial conditions" dates back to the 1970s). This mathematical model, which he explicitly says could be false, nevertheless, attempts to make the world intelligible; the 1950 imitation tends for its part to trick the observer (and will thus be considered at the origin of classical artificial intelligence).

5.2.4 *Rules and the algorithm*

Computerized simulation transforms all physical evolution into an elaboration of digital information. Particularly, the simulation of a geodesic in a discrete universe should make a digital computation correspond to a trajectory, and make the conservation of information correspond to laws of conservation (energy, movement, ...). Any state or property, in short, any physical quantity, as a determination of objects (in the sense of D.1), are in fact encoded by digital information; the quantity of movement is encoded using 0s/1s, just as is the intensity of a field or mass, and their evolution is a calculus approximated by these 0s/1s. Is this encoding "conservational" (does it preserve that which is important)? If a physical trajectory is a geodesic, which geodesic do we associate with the calculus in its digital universe, which symmetry breaking do we associate with rounding?

Let's begin by recalling the generality of the geodesic principles in physics at the center of this science since Copernicus, Kepler, and Galileo. As we observed in Chapter 4 and as we recall again here: any fundamental law of physics is the expression of a geodesic principle applied within the appropriate space. The work of the physicist, who organizes and, by that, makes the phenomena intelligible, consists in a good measure of the search for this space (conceptual, or mathematical) and for its relevant metric.

This approach leads us to understand the slide of meaning around the concept of law, which intends to justify the identification of mathematical modeling with computational imitation. The notion of "law" has a social origin: law is normative to human behavior. The transferal of the concept as it is to physics corresponds to an ordinary metaphysics: an *a priori* (divine, if possible), which would dictate the world's laws of evolution. Matter would then conform itself to this pre-existing and normative ontology (as mathematical laws of a Platonic universe, for instance). On the other hand, the comprehension of the notion of law as an *explication of the regularities and criticalities of a landscape*, with its mountain passes, valleys, and peaks, its geodesics, inverses this and highlights the transcendental constitution at the center of any construction of knowledge. Mathematics, the tool of formal determination, then elaborates itself upon the phenomenal veil at the interface between us and reality, the reality which, of course, causes friction and canalizes the cognitive act, but which is also organized by this same act. Laws are not "already there," but are co-constituted in the intrication between ourselves and the world: the discernability of geodesics, as formal determinations within the framework of a network of interactions, is its main result. And their mathematical processing coincides with the beginning of modern science. The normativity of physical law then becomes only cognitive (in order to construct knowledge), and not an ontology. The different forms of formal determination (laws) propose different causal regimes, the ulterior tools for intelligibility.

The identification between algorithm and law causes us to make a backwards step: the algorithm is normative for the machine, for its calculi, exactly as God's law governs any trajectory. The machine would not know where to go; it would be static, without its primary motor, the program. Once again, the myth of the computer-universe (the genome, evolution, the brain ... all governed by algorithms) consists of a metaphysics and a notion of determination which precedes the science of the XXth century and for good historical reasons: the re-centering of the foundations of mathematics upon a philosophy of the arithmetic absolute, at the fringes of the time's great scientific turning points, as we have mentioned a few times.

This way of understanding law should highlight the very first difficulty for the computational simulation of a physical trajectory by means of a calculus. Physical law and algorithm therefore do not coincide: they do not have the same epistemological status. Law is also not an algorithm for another reason we have already evoked: the formal determination, as mathematical explanation of laws, does not imply the predictability of physical evolution. On the other hand, any algorithm, implemented within a discrete state machine, generates a predictable calculus, at least thanks to the "symmetry" by repetition within time, which we mentioned earlier (identical iteration always being possible for a sequential computer).

However, we absolutely need digital simulation, a tool which is indispensable today for any sort of construction of scientific knowledge: by highlighting the differences, outside of any computational myth (the world would be like a computer), we aim to better identify that which is doable, and afterwards to do better, in terms of simulation-imitation. Let's attempt then to understand the evolution of a calculus in physical terms.

In the case of an isolated computer – a sequential machine – we remain within a Newtonian framework: the absoluteness of the clock and of the access to a database are its essential characteristics. The situation is more complex with modern networks: the distribution of machines in physical space and the ensuing relational time changes the situation. Certain aspects of the absoluteness of Turing machines are thrown into question (we discuss this in detail in Longo (2007) and, more technically, in Aceto et al. (2003)). However, the *exactitude* of the discrete database subsists, as well as the issue of rounding, of course.

In both cases, of sequentiality and concurrence, we may nevertheless understand calculus as a geodesic within *the space (pre)determined by the program* (more specifically: by the programming environment, or all aspects of software – operating system, compilers and interpreters, programs, ...). Shortly, while accepting the divine *a priori* of the programmer who establishes, beforehand, the rules of the game, the notion of "following a geodesic" would be defined by *following the rule* correctly. The hardware or software bug would then be the fluctuation or perturbation that causes the evolution to derail. However, this type of bug is not integral to the theory; it is not inherent to it, contrarily to the theory of dynamic systems which integrates the notion of sensitivity to the conditions at the edges as well as measurement by interval. Moreover, hardware bugs and logical errors are very rare (therefore, statistically very different to the variation due to approximation within a dynamic); they are to be avoided and, in principle, are avoidable (or they may belong to another phenomenal level, which is far from being integrated within the mathematics of effective calculus: quantum physics).

We are left with the issue of rounding, which is inherent to calculus. The use of rounding, today, can be very dynamic and mobile: one can aim to a desired approximation and the end of a calculus and increase beforehand the available decimals, up to hundreds, to end within the targeted interval, if possible. The modern approach to analysis by intervals provides a powerful theoretical framework for these processes (Edalat, 1997). Of course, the speed of the calculi is inversely proportional to the improvement of the approximation. An excess of the latter may prevent us from following any dynamic long enough.

This being said, this bound, the rounding, constitutive of the arithmetization of the world (a quite necessary arithmetization if one wants digital machines to perform calculi and therefore to participate in science today), modifies the causal regime and the symmetries correlated to it, as we hinted and keep demonstrating.

Firstly, let's clear up a possible confusion: the interval inherent to classical physical measurement and quantum incertitude have nothing to do with digital approximation. First, measurement as interval is a physical principle, a classical one; it is not a "practical" issue: thermal fluctuation, at least, is always present above absolute zero, by principle. And, as we have already and often observed, the fluctuation or perturbation below the observable amplitude participate in the evolution of a dynamic system, which is somewhat unstable, because it can break the symmetries of the evolution, and thus be one of the causes of a specific trajectory. In computer science, a bug which manifests below the rounding is without effect. Afterward, the analogy sometimes made – naively – between digital discretization and that of elements of "length" (time and space) induced by the Plank constant, h, is not relevant. The non-separability, the non-locality, the essential indetermination of which we speak in quantum physics are almost the opposite of the certitude of the little boxes, well *localized* and stable, well *separated* by predicates (the memory addresses), within which is distributed the digital universe.

So we now face the principal issue: rounding entails a loss of information, at each step of the calculus. It can be associated with the irreversible growth of a form of entropy, defined as negative information (neg-information) So if one encodes all determinations, formal and objective, of a physical object and process, all properties and states, in the form of digital information, the elaboration of the latter, the digital calculus, will follow a geodesic which is, normally, perturbed at each step by a loss of information. This perturbation does not correspond to any phenomenon intrinsic to the process we intend to simulate: the loss of information is not, generally, the encoding of the change in objective determination. It is a new type of symmetry breaking. Will this influence the proximity of the virtual reality to the physical phenomenon? Will it influence the quality of the imitation?

As we have already observed, arithmetic approximation does not affect the simulation of a linear or Laplacian process: just as the approximation of the measurement, the rounding, does not remove the computational geodesic (the following of the rule) from the physical geodesic. The initial loss of information is preserved, it remains of the same order of magnitude or it increases in a controlled way (technically: the extremes of the approximation intervals are preserved). This is not the case for non-linear cases. Let's consider, for illustrative purposes, one of the simplest dynamics, one that is well-known and one-dimensional: the discrete logistic equation,

$$x_{n+1} = kx_n(1 - x_n)$$

For $2 \leq k \leq 4$, this equation formally defines a sequence $\{x_i\}$ of real numbers, between 0 and 1 (a "time discrete" trajectory within a continuous space). Particularly, for $k = 4$, it generates chaotic trajectories (sensitive to the initial conditions, dense in $[0, 1]$, with an infinity of periodic points, ...). Can we approximate any sequence of real numbers thus generated by a digital computer? Out of the question, at least in what concerns an initial value of x_o taken from a set of measure 1 (that is, for almost any real value in $[0, 1]$). Even if we choose a x_o that can be represented exactly by a computer, at the first rounding in the course of the calculus, the digital sequence and the continuous sequence will begin to diverge. By improving the approximation/rounding of 10^{-14} to 10^{-15}, for instance, after approximately 40 iterations, the distance between the two sequences will start to oscillate between 0 and 1 (the greatest possible distance). Likewise if, with a rounding of 10^{-15}, we begin using values that differ from 10^{-14} (of course, if we want to, nothing would prevent us from restarting the digital machine upon the exact same values and from calculating, with the same rounding, exactly the same discrete trajectory, ...). The technical problem can be summed up by the observation that the dynamic is a "shuffling" one: the boundaries of the interval are not preserved.

We therefore may not, in general, approximate, with the machine, a continuous trajectory; however, we may do the opposite. In fact, all that can be proved, in dynamic (metric) contexts we will not specify here, is the following "pursuit" lemma (see the *Shadowing Lemma* (Pilyugin, 1999): notice the order of the logical quantification):

For any x_o and δ there is a ε such as that for any trajectory f, ε-approximated (or with a rounding-off not greater than ε at each step), there exists a continuous one, g, such as g approaches f by a difference of at most δ at each step.

Even when considering the lucky case where we have $\delta = \varepsilon$ (this is possible in certain cases), it comes to say that, globally, your digital sequences are not so "wild": they can be approximated by a continuous sequence, or ... there are so many continuous trajectories that, if we take a discrete one, one can easily find a continuous one, which is close to it. Thus, the image of an attractor on the screen provides qualitatively correct information: the digital trajectories are approximated by trajectories of continuous dynamics (determined by equations). But the reverse is not true: that is, it is not true in general, for a trajectory given by analytical means, that the computer may always approximate it. Different versions of the pursuit lemma apply to sufficiently regular chaotic systems. However, many dynamic systems do not even satisfy weak forms of this lemma (see Sauer, 2003). This signifies the existence of initial values and intervals such that, within these intervals, any rounding and any other initial value cause any continuous sequence to diverge from the given discrete sequence.

What happens, in the terms of our approach, which is geometric in nature? To understand this in detail, it would be necessary to refer to the technical analysis which the authors are also developing. In this work of reflection, which nevertheless guides the mathematical and computational analysis, let's try to see it in a very informal manner. The first difficulty resides in the necessity to place oneself within the appropriate space, in order to better understand. In short, it is necessary to analyze the evolution of a system, such as for instance the discrete logistic function, within a space where the notion of neighborhood, between real numbers, corresponds to digital approximation. A space which provides for such a metric is called a "Cantor space." In this space, which we will not define here, two real points are close if and only if they have close binary or decimal representations (for instance, 0.199999... to infinity and 0.2 are very far from one another in the Cantor space, whereas they are identical with regard to the usual real line, thus posing numerous problems from the computational standpoint, when we try to operate upon their approximations).

We then see that at each of the digital calculus' iterations, the rounding induces a loss of information corresponding to the deterioration of the approximation around the point of the trajectory. If we measure the phenomenon in terms of isotropy of space (the points within this widening neighborhood are "indistinguishable," in a manner of speaking), this "gray" zone, of isotropy, grows, thus augmenting the symmetries of space. A notion of entropy as negative information also enables us to grasp this change in symmetries, as a loss of information. Now, all that we have in the machine is encoded information. Independently of what it encodes, the physical object's formally determined properties or states, all is in the form of digital information. So the objective determination, which is given by the preservation of the theoretical symmetries, radically changes: we are facing a change in symmetry that does not model a component of the evolution of the natural phenomenon, because it depends only upon the discrete structure of the simulation universe and upon the imitation of the formal physical determination by algorithms (or, when we make a philosophy of it, of the epistemological identification of *law* with *algorithm*).

So there is, in terms of symmetry, the explanation of the causal regime of which we were speaking. The discretization, in fact the organization of the world by means of discrete mathematics proposes a causal regime (in this case, an evolution of symmetries), which is different than that which is proposed by continuous mathematics. It is not an issue of finitary translations of the same physical world, but of scientific construction, because this world is itself co-constituted by our formal and objective determinations. When they change, its organization and its intelligibility also change. Once more, this does not imply that the world is continuous "in itself": we are only observing that, since Newton, Leibniz, Riemann, Poincaré, ... we have organized and made intelligible some physical phenomena by means of historical notions of continuity and of limit. If we want to do without them, causal organization and intelligibility will be altered.

Another issue would also merit to be detailed, but we will leave it for latter work. Singularities in modern physics play an essential role. We know for instance of shock situations, in non-linear systems, where the digital calculus does not come even remotely close to the critical situation. We have the continuous description; the mathematics is clear, explicative, organizing for the physical phenomenon, we understand qualitatively, but the numerical calculi chaotically revolve around the singularity, without coming close to it. In fact, the current notions of limit and of singular point, which are absolutely necessary to the analysis of the phase changes, the shocks, in order to even speak of renormalization processes in physics, are not always coherently approximable. The loss of symmetries and the change of a correlated causal regime constitute our way of understanding this problem, which is specific to the digitalization of phenomena, without referring to Laplacian myths and computational metaphysics. Computer science, a science which is now mature, deserves, from an epistemological and mathematical standpoint, more attention and an internal view which is able to assume the force and the limits of its own methods.

5.3 Causalities in Biology

While focusing now on biology, we will not address the discussions concerning the biological levels of organization, the intertwined hierarchies, the crossed causalities, the ago-antagonistic effects, the variabilities within phenomena, the autopoietic processes, which we find in biology. Of course, all these properties will remain part of the backdrop of the approach, which we propose here, but the approach will be more conceptual or, better, "schematic" (in the geometric sense of simple schemas or diagrams) than thoroughly theoretical or descriptive: it will seek rather to contemplate a framework of representation enabling us to extricate heuristic categories of thought than to account for the effective phenomenality of life. Indeed, the "relationships," which we highlight, by means of very abstract little patterns, do not necessarily correspond to "material relationships" or to physical configurations; they are only organizing structures of thought, which should aid the comprehension of phenomena, by proposing a conceptual framework. Moreover, is $\mathbf{F} = m\mathbf{a}$ – at a much more elaborate and mathematized level indeed – not a correlation which organizes a phenomenon by making it intelligible? Let's also recall that this equation has been preceded by the general concept of inertia, or even, way before Galileo, by cosmological speculations and concepts as eminently philosophical as profound (see, for instance, the remarks of Giordano Bruno in "*L'infinito universo e mondi*," 1584). The physical intelligibility specific to this equation can be the object of highly differing conceptual "readings": it may no longer be primitive, but derived (from the Hamiltonian, from the Lagrangian, as we have mentioned in the first part), where it may be correlated to distinct symmetry breaking, as we have also seen.

So, as we insist on emphasizing from the onset, our approach remains very speculative here: it is for us the beginning of an attempt at a conceptual categorization and schematization, which seeks to open new venues without being sure of their outcomes and which will require, in order to be continued, more discussion with biologists and the sanction of a certain fecundity in the quest for a greater understanding of living phenomena. Of course, we will remain within the framework, which we have determined for ourselves, that is in the geometric terms of symmetries as an analysis of physical causality; yet, we will *also* take into account here aspects which are specific to biology, related to forms of teleonomies or of anticipation and which we have already summed up with the concept of "contingent finality" (see the next section).

5.3.1 *Basic representation*

Let's consider the dynamic functioning of a centrifugal governor (or Watt governor: two weights are lifted in rotation by pressure, making a valve open so as to lower the pressure ...). This functioning is completely determined by the data, obtained afterwards, concerning the initial conditions and the physical laws. In this sense, its "behavior," though being well regulated and leading to a dynamic equilibrium, is determined in a univocal and oriented way (geodesics within a well – and pre-established phase space).

In the case of living phenomena, the behavior (and functioning) of an organism does not appear to be determined in the same way. What appears to be determined like this in a more or less rigid way (within a given domain, that is compatible with the organism's survival), is what we could call the aim of the functionings and behaviors, the functions to be fulfilled in order to ensure homeostasis (or, better, homeorhesis, as any equilibrium is dynamically, not statically preserved, in biology). Yet, what is not determined in such a manner, are on the one hand the possible ways to achieve this and on the other hand the adaptations and modulations, which would ensure the achievement of the functions and behaviors. Specifically, to put it in the language of physics, there are not only phase changes but also, as we have seen in Chapter 3, changes in phase spaces, that is, observables and relevant variables.

We know that the mathematical and equational formalization of this situation (as took place for physics and as we hope, in time, could be the case for life sciences inasmuch as the adequate mathematics would be elaborated) is confronted with profound difficulties, sometimes even difficulties of principle, which the numerous and successive attempts at modeling have encountered. Also, before any reiterated attempt in this regard, it appears to be necessary to try to illustrate and to represent – in this case by means of schematics, the first abstract conceptual stage – that which appears to characterize these modes of functioning and that which we could call the "finalities" which interpret them, in the sense in which Monod could speak of a *teleonomy* of life. These finalities, of course, are neither necessary nor absolute; they rather participate in our view of living phenomena and, most of all, they are contingent, as they are specific to living matter and relative to its contexts. In short, they could not be present (no life, no specific specie or individual); they are relevant to various levels of organization and to their correlations, or to their intertwining and looping, particularly in the form of integration and regulation (see Bailly and Longo, 2003 and 2009).

To make intelligible the notion of contingent finality, we will try to organize into networks the interactions between the "material structures" and "functions" of living matter. It is at this level in fact that teleonomy manifests: for instance, when an organic structure appears to be finalized in relation to a certain function.

We therefore propose a conceptual framework, to organize knowledge by means of our recourse – temporarily at least – to a description of this sort (for a better, active, understanding, the reader may easily draw the simple intended diagrams):

(1) We have a target set, constituted of several domains – the target domains, which may or may not overlap – corresponding to the functions to be ensured for the maintenance and the perdurance of the organism and its species.
(2) We have a source set constituted of all the organic possibilities likely to be mobilized to this end (biochemical reactions, transport agents, etc.), also represented by this set's domains (source domains).
(3) We have a set of arrows, originating in the source domains to reach to the target domains (these arrows correspond to the orientations and functioning modes aiming to ensure the functions) and which presents the following particularities:

(i) Any target domain is reached by at least one arrow; usually, several source domains are at the origin of the arrows reaching the same target domain. An example of this situation is the conjunction of "oxygen metabolism" and of "glucose metabolism" to ensure the maintenance of muscular tissue such as the heart; in this case both source domains are respectively defined by the chemical reactions related to the specific energetic sources (availability of glucose) and by the cellular assimilation processes of the intake of oxygen obtained through breathing (availability of oxygen), the arrows corresponding for their part to the various transport and transformation systems which enable the effective transferals from sources to targets.

(ii) The arrows pointing to the same target domain are endowed with different widths depending on the prevalence of the usual modes of functioning (in the preceding example, we would have, in the normal case, an arrow width for "oxygen metabolism" that is much greater than that of the "glucose metabolism." In the case where a dominant mode of functioning would fail (pathology), the corresponding arrow could narrow down to the benefit of another whose initial width was smaller (and this, without necessarily reaching the width of the first one: functional weakening all the while attempting to preserve the function): this mechanism would correspond to a property of *plasticity*.

(iii) The arrows stemming from the same source domain and extending towards several target domains exist, but may be relatively rare in adult homeostatic (-rhersic) functioning. They refer mainly to potentialities, preceding ulterior actualizations or differentiations (*cf.* stem cells, for instance), or to other possibilities of plasticity (cerebral, for instance). On the other hand, in the case of the representation of a *genesis* (embryogenesis, namely), these arrows are dominant and play an essential role in the representation of the organic differentiations from totipotent eggs or from pluripotent stem cells. There is, therefore, a dynamic for the "topology" and the width of the arrows over the course of development to reach the adult situation.

It could be interesting and enlightening to note here that the joining of the characteristics (i) and (iii), for the arrows, corresponds rather well to the very important concept of *degeneracy* such as it has been introduced by Edelman and Tononi (2000) with regard to cerebral functioning (that non-isomorphic structures may participate in the same functionality and that a given structure may participate in several of these functionalities). This concept returns to and generalizes that of *redundancy*, but while differing somewhat from it (computer redundancy, for example, is achieved by iteration of identical components). In this perspective, we could qualify the described situation by the characteristics in (i) of *"systemic" degeneracy*, (the same system participating in distinct functions) and the characteristics in (iii) of *"functional" degeneracy* (non-isomorphic systems participating in the same function). Degeneracy (perhaps an unfortunate name) is present everywhere in biology, from DNA to the brain. Different segments of DNA may be used by the cell to produce the same protein; conversely, a given segment of DNA may be used to produce different proteins ("alternative splicing," see Longo and Tendero (2007) for a discussion and references). As for the brain, Edelman and Tononi thematize and better specify by this notion the well known, and often surprising, plasticity of this organ.

Let's also point out right away that the concepts of "source domain" and "target domain" do not necessarily refer to "absolute" categorizations, but are relative to a given functionality (or to a set of functionalities): a target domain for a functionality can very well operate as a source domain for another[6] on the same level of organization or between levels, thus the many possible intertwinings.

Notice on the other hand that, in this approach, the environmental, feedback or adaptation effects can be represented by variations in the widths of the arrows ("metric" aspect), or that the fundamental changes would rather correspond to changes in the structuring of the set of arrows ("topological" aspect). Moreover, pathology is likely to occur (in order of "seriousness"):

- either with a variation in the width of the arrows,
- either with the disappearance of certain arrows (without, nevertheless, a target domain being no longer concerned at all),
- or with the disappearance of the source domains (in this case, grafts and prostheses can play an "artificial" regulatory role).

We may consider that the disappearance of the target domains corresponds at best to a mutation, and in the worst of cases, to death.

To give a few "systematic" examples of the functioning thus represented, we can propose the following triplets (by starting with the source domains, then arrows – in fact corresponding to functions – and finishing with target domains):

- Vascular system / circulation (transport) / local essentials (nutrients, oxygen, etc.);
- Respiratory system / breathing / oxygenation;

---

[6]For instance, the putting into effect of ionic equilibrium processes may constitute a source domain for the functioning of the target domain represented by a cell, itself constituting a source domain for the good functioning of the tissues in which it participates, good functioning representing one of its target domains. It would go likewise for cerebral functioning, for example, as target domain for an oxygenation and as source domain for a control or behavior.

- Nervous system / information, command / adaptation, initiative;
- Genes / expression / proteins, regulation;
- Mitochondria / biochemical reactions / energy produced;
- Digestive system / digestion and transport / metabolism;
- Immune system / reconnaissance / tissue identity, struggle against aggressions

5.3.2 On contingent finality

Based upon these considerations, we can propose to call *contingent finality* the abstract structure formed,

(1) by the triplet {source domain, arrows, target domains},
(2) endowed with the "measurement" constituted by the set of real numbers E, of the widths of the n arrows: $E = \{e_1, e_2, \ldots, e_n\}$,
(3) ensuring a *structural stability* for these characterizations. We mean, by such structural stability, the conservation of the target domains in the sense that there will always be at least one arrow for which the width is non null and which points to these targets, regardless of the source domains.

Let's go back to the preceding example and let's attempt to compare a normal state to a pathological state. In the normal state, the "oxygen metabolism" arrow has a width of e_{O1} and the "glucose metabolism" arrow has a width of e_{G1}, with $e_{O1} \gg e_{G1}$ and $e_{O1} + e_{G1} = e_1$. The establishment of the pathological state is translated by a narrowing of the "oxygen" arrow and the widening of the "glucose" arrow; finally, we have
$e_{O2} < e_{O1}$,
$e_{G2} > e_{G1}$,
$e_{O2} + e_{G2} = e_2 < e_1$.

The fact that the arrows do not cancel each other out and that the target domain remains translates a partial *plasticity*, whereas the decrease of the total width, on the one hand, and the internal rebalancing of the widths, on the other, demonstrates the pathological character. The influence of these two factors (total width and respective widths) could indicate that total plasticity (in the sense where we would finally have $e_2 = e_1$) does not, however, restore a completely "normal" situation.

From a much more general point of view, we will notice that – as we have already highlighted through the approach by levels of organization, previously considered – the same structure of "contingent finalization" thus defined, replicates itself at various levels of organization of biolons (cell, organism, species)[7], even if the characterizations (triplets and measurements) may differ in their specific content, according to the level. This structural likeness is doubtlessly the result of a certain form of equivalence of the objective complexities associated with these levels, as we have already noted (Bailly and Longo, 2003).[8]

5.3.3 "Causal" dynamics: Development, maturity, aging, death

Let's note here that if we accept the schema we have just discussed, it proves likely to represent, thanks to the topological and "metric" plasticity it is able to demonstrate, the great dynamic processes of which life can be the locus: the beginning of development is characterized by the prevalence of arrows which stem from a source domain to point towards several target domains, which they contribute to constitute (differentiation of tissues and of anatomical and physical systems). As the process unfolds and at the same time as the number and structure of the target domains stabilizes, these arrows narrow down (some may even disappear) at the same time as the arrows originating in several source domains ending at the same target domain (functional aims) start to prevail. The set stabilizes once again

---

[7] Synthetically, we call a living entity a *biolon* (a cell, an individual, animal or plant or an entire species). Biolons are composed of *orgons* (the organelles of a cell, the organs of an individual, the organized populations of a species). This terminological unification is justified by the uniformity of concepts, of mathematical tools, with which one can address the three levels brought together under the same name (see (Bailly *et al.*, 1993) and (Bailly and Longo, 2003), where the notion of extended criticality is just hinted at).

[8] Let's recall that, according to our analysis in (Bailly and Longo, 2003) where we distinguished between objective complexity and epistemic complexity in biology, also providing examples, the elements of living matter, which are biolons, present an objective complexity which may be considered as being infinite, with respect to any physical measure (crossing of the essential level of organization which enables us to pass from inert to life). From this point of view, and still with regard to physical complexity, the objective complexity of biological objects is comparable regardless of what these biolon-type objects are (the living cell presents an objective complexity almost identical to that of an organism such as a mammal). What is modified along life's scale of complexity is epistemic complexity, related to the enriching structure of phenotypes, to the increasing, along evolution, levels of organization, their intertwining, the proliferation of structures and functions, the conditions of description, etc (see Bailly and Longo (2009) for an analysis of complexity in evolution).

following development, the period of maturity.

Once the stability of the maturation has beenachieved, the topology maintains itself "on the whole" and aging manifests itself mainly in a "metric" fashion (by the variation of the *measurement* of the narrowing of arrows). It is even possible that in borderline cases one may witness disappearances of arrows by cancellation of their widths, amounting, beyond metrics, to tapping onto the topological structure of the schema. And finally, to represent the death of an organism, we may agree, as suggested above, that it manifests itself by the disappearance of one or more target domains (corresponding to vital functions), in that there are no more arrows pointing towards them.

We will note that if most of an individual's target domains are oriented towards the individual's perdurance, at least one of them, corresponding to the reproductive function – is likely to produce a new source domain (child cell, fertilized egg) as the origin of the reiteration of the process for a new individual. It is the set formed by the abstract joining of this particular target domain to the newly produced source domains which may constitute – at a different level – the source domain at the origin of the genesis of individuals of the level thus considered (organism for cells, species for individuals).

From the standpoint of an attempt at a more precise "phenomenal" identification of the characteristics we have just introduced abstractly, we may consider that in the initial "transitory regime" (time of genesis) the source domains are principally constituted by biolons(embryonic cells, individual organisms, species, as defined in (Bailly *et al.*, 1993) and further analyzed in chapter 6), the target domains being mainly constituted by orgons (cell's organelles, organs and tissues to be set). The arrows corresponding for their part to the phenomena of differentiation, of migration and of structuration, whereas, conversely, in the "stationary regime" (adult organism), the source domains are mainly constituted by the constitutive orgons, whilst the target domains would be constituted by the variety of vital functions ensuring the organism's maintenance and autonomy, the arrows corresponding this time to the biochemical and physical processes of these functions (integration and regulation). Such an approach would enable us to propose a sort of "temporalized" schema of biological functioning.

Could we refine the analysis by taking more precisely into account the nature of the fluxes which link source and target domains to one another? Namely by putting the distinction between energy and information to work? In a first approach, it appears legitimate to consider that the fluxes in the source/target direction mainly have an energetic character (transport of matter or energy), responding to fluxes which are mainly of information (gradients, divergences from the dynamic equilibrium) going in the target/source direction. The arrows are then supposed to integrate and represent both types of fluxes, their width alterable in the event of a failing of either the correlated "informative" or of the "energetic" character (we could, in a first approximation, take as parameter the product of these two types of fluxes, for instance[9]). Let's try to take an example at one of the most elementary levels, that of the cell: in this case, a particular source domain can be associated with the functioning of ionic channels enabling the ions to cross the cellular membrane and a corresponding source domain would be the stationarity (dynamic equilibrium) of the cell's internal ionic state (homeostasis – homeorhesis) which enables it to function optimally. The arrows would then correspond to both "fluxes": on the one hand, the "information" flux, which would be generated by a difference in the internal ionic concentration with regard to the stationary state (gradient, difference in osmotic pressure, electric field, ...) and which would lead for instance to the opening of certain channels, and on the other hand the concomitant flux of matter (these same ions) coming from the outside in view of re-establishing homeostasis and entering through these channels.

Notice that, from a standpoint analogical to this stage, considering these two aspects (matter/energy and information) highly resembles the thermodynamic situation where the definition of free energy (of which the variations govern the system's evolution) entails the intervention on the one hand of an enthalpy (or internal energy) and on the other hand of an entropy, these two magnitudes being associated by means of temperature.

5.3.4 Invariants of causal reduction in biology

As in the case of physics, we may question the invariants of causal reduction (if they exist) specific to life and their relationships with what could constitute, in the field of biology, the determinations associated with the

---

[9] If E is the matter-energy flux extending from the source domain to the target domain in order to "respond" to the information flux ("request") going from target to source domain "within" a given arrow, we could take as one of the parameters of functioning – which participates in the width of this arrow – the product E x F. Thus, the failing of a flux in one or the other direction would translate as a diminution of this product, corresponding to a decrease of the width of the arrow and thus expressing an alteration of the functional process summarized by this arrow.

symmetries which we have encountered in physics.

As we have emphasized many times (see Chapters 3 and 4), it very well appears that these biological invariants exist indeed and that they are constituted by sets of pure, numerical invariants (and not dimensional invariants as in physics). It also appears that the determinations which enframe, modulate, and actualize them are now rules of "scaling" in function, let's say, of the size or mass of the organisms (sorts of dilatation or scale symmetries), see also Schmidt-Nielsen (1984).

Thus, let's recall that the average life-spans of the set of organisms do appear to scale as a power $1/4$ of their masses and their metabolism as a power $3/4$ of these masses (Peters, 1983). Likewise, on a level that is somewhat different but which is in relationship with these properties, we would recall here that mammals are characterized by an invariant average number of heartbeats or breaths (of the order of 10^9 heartbeats or of $2.5x10^8$ breats over the course of an average life), these numbers are equivalent to frequencies (or periods) – dimensional magnitudes, this time – which are submitted to these rules of scaling as a function of the average mass of the individuals of the considered species (for instance with a power of $-1/4$ of mass for the frequencies).

But such characteristics of invariance do not manifest solely at the high level of the biological functions of evolved organisms; they can also be found at much more elementary levels such as those of cellular metabolic networks[10] (Ricard, 2003; Jeong et al., 2000), of which the diameter[11] remains invariant along the phylogenetic tree and of which the connectivity distribution, over at least 43 organisms belonging to the three categories of life forms,[12] presents the same characteristic power (2.2 approximately). As emphasized by Ricard, such an invariance of the network's diameter implies that the degree of connection of nodes increases with the number of these nodes, that is, with the number of stages likely to connect them. Here again, one will notice that it is a case of numeric and not dimensional invariants.

---

[10] We know that a network is a graph formed by nodes, which are connected according to certain rules. In the case of metabolic networks, these may be constituted by metabolites or by enzymatic reactions and the connection links representing the mutual biochemical interactions.

[11] The diameter of a metabolic network is defined by the average of the shortest paths (in terms of steps), leading from one of the network's nodes to another.

[12] That is, archaebacteria, bacteria, eukaryotes.

5.3.5 *A few comments and comparisons with physics*

We see that viewed from this angle, the causality, which seems to manifest in living phenomena, presents similar traits as well as different traits with regard to those which we have noted in the case of physics, which only addresses inert matter, which we theoretically care to distinguish from the "living state of matter." Material and efficient causality are manifestly present in biology – this having doubtlessly favored the idea of a possible physicalistic reduction – although in a much less rigorous and structured way than for physics (namely in connection with the plasticity and adaptability capacities). The formal determinations are relatively weakly represented there, despite the advances made in terms of various local models (we have mentioned the metabolic networks, but at a different level, we could evoke population dynamics, for instance, or the transport properties close to the fluid dynamics). Likewise, the essential objective determinations do not really appear to have been extricated, despite the observation of variegated properties of symmetry – or of symmetry breaking – (over the course of development or in certain anatomies, for instance) and the identification of certain digital invariants. On the other hand, the dimension of "final causality" (to use old categorizations) or of "contingent finality" appears to play here a role which is rather important and unknown to physics. As if the fact of finding itself in an extended critical state (therefore potentially highly unstable, although necessary to an elaborate organization) could be compensated for the structural (momentary) stabilization of living matter only with the introduction of these factors of teleonomy/anticipation, which appear to characterize it.

5.4 Synthesis and Conclusion

We have attempted to briefly characterize the different aspects of physical causality such as they may appear and are analyzed through contemporary theories. We have emphasized the fact that symmetries and invariances constitute determinations which are even deeper than those which manifest causal laws in that they present themselves, in a way, as the conditions of possibility for the latter and as frames of reference to which they must conform.

We have also sketched out an analysis of the causality internal to the systems of effective computability, of which the symmetries and invariances obey a specific regime, rooted in the discrete arithmetic structure of databases and algorithms. The intelligibility structure proposed by these methods differ by that which is inherent, on one side, to the geometry and mathematics of the phenomenal continuum, particularly by the difference between the (modern) notion of physical law, to which we refer, and, on the other side, to the notion of algorithm. The consequences of these two aspects can be measured in terms of different causal regimes, following differences (breaking) in symmetry. Iteration, as a particular symmetry in time, is also mentioned as one of the characteristics of digital simulation, in fact as one of the strong points of computational imitation (and a starting point for effective recursion, as a mathematical theory). It is also at the center of the particular status of predictability, even in the case of the computer implementation of highly unstable non-linear systems, because the possibility of identically iterating a process (or of accelerating a simulation) is a form of prediction. Iterability, that is, digital calculi, also enables us to grasp the difference between the randomness of the theory of algorithmic information and the randomness of physical processes of the critical and quantum type. In the case of algorithms, randomness coincides with incompressibility. On the other hand, in the first of the physical cases (deterministic, dynamic, and thermodynamic systems), it is of an epistemic nature and it implies the non-iterability of processes; in the second (quantum physics), it is intrinsic to the theory (it is part of the objective determination, see chapter 7). These two last cases are incompatible with the individual iterability of an individual process (which is typical of algorithmics); although it may have a statistical iterability, as in quantum mechanics.

We have then attempted to widen the causal problematic to encompass life by taking into account its specific character through what appears as a sort of finalization of its functioning, which we have attempted to conceptually systematize. This led us to refer to specific concepts, such as that of "contingent finality," and to propose new representations (topological or metrical) in order to attempt to account for it in a more or less operational fashion. Finally, we have evoked the possibility of extricating that which – through numerical constants and scaling properties – could be considered as invariants of causal reduction specific to life.

If the considerations relative to physics and to calculus base themselves upon well elaborated and mathematized theories, enabling us to refer to a *corpus* that is almost completely objectivized and thus lending itself particularly well to a thorough conceptual and epistemological analysis, moreover, situating itself within the framework of a well-established tradition, then it seems that for biology the situation is much more fragile in this regard. Also, with regard to life, the analyses we propose are of a much more speculative nature and require with greater necessity the theoretical and conceptual sanctions concerning experimental practices in this field. More so that the causal representations, in the case of life, if we seek to detail them, must take into account multiple interactions, which present themselves simultaneously all the while remaining of a highly different nature, whereas in physics, for example, the interactions may in many cases be sufficiently decoupled from one another for us to be able to approach and study them separately; even if it implies, in a second stage, to seek the conditions and procedures for their unification. In other words and in particular, in biology, the causal representations must take account of massive retroactions, which prevent them most often from partitioning systems into weakly coupled sub-systems in order to facilitate analysis, as is done in physics, as well as to consider holistic teleonomies, according to which the local organization is dependent upon the global structure and reorganizes itself according to the necessities of optimization or of perdurance of this structure according to criteria, which are still not well known. We know, for example, that the genetic constraints themselves manifest "normally" only in adequate extragenomic or environmental frameworks. Moreover, some phenomena, such as apoptosis,[13] apparently in contrast to evolutionary-survival paradigms, prove to be necessary to the *global* viability, as an integral part of "normal" life and even more of the adaptation faculties of organisms in the event of the modification of the exterior environment.

Nevertheless, it appears that one of the points common to these disciplinary fields – and this is what we have wanted to highlight and to emphasize in this text – resides in the fact that the causal analysis, all the while remaining useful and efficient – must now be relativized and henceforth give way, for the purpose of a better comprehension of the theoretical and conceptual structures of these fields, to a more general approach relying much more upon the properties of invariance, of symmetry (and its breaking) and of conservation, as we stressed. These properties underlie the manifestations which we tend to spontaneously (at least since the Renaissance) interpret in terms of objective causal actions. We will maybe see there the trace of a process of conceptual rehabilitation of the "geometrical"

---

[13] *Local* "death" of cells contributing to embryogenesis or growth and reconstruction of tissues; typically, fingers are formed by the death of the cells "in between."

(taken in a very broad sense) in relation to the "arithmetical,"[14] which is not without echoing the most profound preoccupations of this volume.

---

[14] In reference to the debates at the beginning of the XXth century concerning the foundations of mathematics (see Chapters 1 and 2), but all the while emphasizing the fact that the geometrization of physics, for its part, has never ceased to develop, probably explaining why it is in this discipline that symmetries and invariances have acquired a determining explicative and operational status rather early on.

Chapter 6

Extended Criticality: The Physical Singularity of Life Phenomena

Introduction

In this chapter we propose to consider living systems as "coherent critical structures", though extended in space and time, their unity being ensured through global causal relations between levels of organization (integration/regulation). This may be seen as a further contribution to the large amount of work already done on the theme of self-organized criticality. More precisely, our main physical paradigm is provided by the analysis of "phase transitions," as this peculiar form of critical state presents interesting aspects of emergence: the formation of extended correlation lengths and coherence structures, the divergence of some observables with respect to the control parameter(s), etc. However, the "coherent critical structures" which are the main focus of our work cannot be reduced to existing physical approaches, since phase transitions, in physics, are treated as "singular events," corresponding to a specific well-defined value of the control parameter. Whereas our claim is that in the case of living systems, these coherent critical structures are "extended" and organized in such a way that they persist in space and time. The relation of this concept to the theory of autopoiesis, as well as to various forms of teleonomy, often present in biological analyses, will be also discussed.

Our perspective should be clear. Mathematics has a normative role for physics: it organizes reality, particularly since the advent of infinitesimal calculus (the role of differential equations, for instance) and, later on, through the geometrization of physics. It ought to play a similar role, if possible, for other scientific disciplines. Yet, the common idea according to which the *same* mathematical tools, so successful with regard to physics, can play a similar role for biology is often based on a confusion between the *dependency* of biological phenomena upon physical phenomena, and their *reducibility* to such physical phenomena. The fact that the phenomena pertaining to living matter are dependent upon physical structure is a basic presupposition for any monist. However, reduction to physics is a theoretical operation: living matter is reduced to a conceptual organization, which is given and specific to inertia. The first aspect is intimately related to the presupposed foundations of modern science; the second aspect must be implemented and this is generally the case when reference is made to the existing physical theories, which may prove inadequate to achieve this. In this text, and in the view of unifying different facts, we will attempt to better clarify some organizing concepts of a number of biological phenomena. This is, once again, a rather modest attempt at "unification through concepts" – an attempt which should always precede the process of mathematization – by returning to an idea already formulated elsewhere, that of the "criticality" of the living; we "extend" this concept in a way beyond the reach of current physical theories. In our opinion, this concept may prove to be an interpretational key by which to grasp biological phenomenality, in a number of situations.

We insist here on the importance of a preliminary construction of a conceptual frame; it is a hindrance rather than a help to adequate mathematical formulation if lack of sufficient time precipitates the performance of calculations before understanding and conceptualizing.

And yet, as we have seen, mathematics itself is the result of a progressive conceptualization, where crucial notions and structures result from a difficult stabilization of informal practices. In this sense, the perfectly stable conceptual invariants of mathematics result from a "praxis," from an active attempt to understand and to organize the world. Greek geometry, we may say, did not spring from an axiomatic approach, but was the result of a long process of abstraction from action and measurement in physical and sensible space. The recourse to actual infinity by Newton and Leibniz, in the calculus of finite movements, speed and acceleration, is a continuation of several centuries of philosophico-conceptual debates regarding the difference between potential and actual infinity, debates which were even often of a religious nature. Afterward, the unification of sub and supra-lunar physics, Newton's major contribution as it was considered up till then as stemming from distinct ontologies – including by Galileo – was obtained by means of radically new mathematics, the origin of modern infinitesimal calculus. In general, the unification of (apparently) different objective (or epistemic) levels necessitates, in all likelihood, new technical tools or even new conceptual organizations of the object; it is the result of a new synthesis, and not only of a transfer of (mathematical) techniques – even though a good practice of this transfer may eventually help in finding new tools. In this vein, we are attempting here to unify some biological phenomena and structures, by using relatively technical concepts but no explicit mathematics; we are only suggesting by this a possible mathematization. In this trans-disciplinary attempt, we are thus drawing on our experience of "theorization" in the field of mathematics and physics by orienting it towards biology. Our experience differs indeed from the practice of biology, but, nevertheless, makes it possible to distinguish new approaches, which we hope to develop on the basis of a sustained and rigorous dialogue with practitioners in this field.

6.1 On Singularities and Criticality in Physics

6.1.1 *From gas to crystal*

A crystal and a gas are opposite to one another, in what concerns the spatial ordering of their components. In physics, the analysis of the change from one to another, of the phase transition, involves specific concepts: those of critical exponents, of divergence or of the discontinuity of certain quantities related to the system (susceptibilities), of increases in fluctuations, of divergence of correlation lengths, etc.

These notions are associated with the passing through a *critical* state, which radically changes the relevant properties and parameters used to describe the object studied. Particularly, as we have just mentioned, the phase transitions change the correlation lengths, which, in certain cases, may be interpreted as a change in the scope of (possibly causal) relationships or of the "coherence structure" between the elements of the studied physical structure. During the process of change of state, the global structure is completely involved in the behavior of the elements: the local situation depends upon (is correlated to) the global situation. Mathematically, this may be expressed by the fact that this correlation length formally tends towards infinity, for example in the case with first-order transitions, such as a para-/ferromagnetic transition; physically, this means that the determination is global and not local. However, the transition sometimes occurs prior to complete divergence, when the length in question remains finite, like in the case with first-order transitions, such as liquid to solid transitions, at a finite length. In any case, the most precise treatment usually resorts to the theory of the renormalization of measurements and of the parameters at the transition point.[1]

As a matter of fact, in physics, a critical state is the most generally reducible to the critical *point*; renormalization, which enables us to account for the passing from the local to the global, finds itself confined to the value of the parameter for which the transition occurred, which finally makes this renormalization implicitly dependent upon a *single value* for this parameter (putting aside here the dependency as a function of the dimension of the space of embedding or that of the order parameter – in the examples mentioned, the magnetic momentum and the density).

To sum up, in physics, a critical state may be related to a change of phase and to the appearance of critical behaviors of some magnitudes of the system's states – magnetization, density, for example – or of some of its particular characteristics – such as correlation length. It is likely to appear at equilibrium (null fluxes) or far from equilibrium (non-null fluxes). If, in the first case, the mathematico-physical processing is rather well understood and thermodynamics is used for the bridge between the microscopic and macroscopic description; on the other hand, in the second case, far from equilibrium, we are far from having at our disposal theories as satisfactory, for the moment.

It is difficult to gather under a sole characterization the phenomena of the critical type, but a common signature is provided by the divergence tendencies of lengths and times of correlation as well as by a loss of analyticity for functions of the system's state (such as free energy), manifested by the appearance of non-integral critical exponents, occurring in the phase transitions. Another common mathematical aspect is found in the fact that the set of critical points, which are located in the space of control parameters, is generally of null measure. It possibly is a discrete set of points, as observed above, relatively to the evolution of the parameter(s) (the temperature, for instance, see also section 2.2.2). We notice that the critical situation, in the case of the null flux, appears to be atemporal, whereas, far

---

[1] Renormalization describes a change in measurement and of object, obtained by integrating the new classes of interaction due to the transition. Its properties stem from the fact that at a critical transition, the passing to the infinite limit of the correlation length produces an invariance of scale for the system and a fixed point from its dynamic (see Delamotte, 2004, for an introduction).

from equilibrium and because of fluxes, also when the system remains in a stationary state, there exists a natural time scale.

6.1.2 From the local to the global

Let's return for a moment to certain characteristics which we wish to highlight and which will enable us to better apprehend the relationships between "local" and "global" as they continue to impose themselves upon the sphere of living phenomena from the moment that relationships of whole to parts are involved, *via* the processes of regulation and integration, among others. Thus, as we have just seen, during critical transitions (change of phases), several characteristics are manifest in the change from the local to the global in relation to the relevance of the described objects and of their interaction:

- the divergence of the correlation length, which is a global statistical property, different but based also on local interactions – e.g. spin coupling;
- the appearance of an order, possibly measured by an order parameter which becomes non-null, a symmetry breaking;
- the appearance of critical exponents associated with discontinuities or with divergences (susceptibilities);
- the sudden non-analyticity of free energy, from the thermodynamic point of view.

In the case of the critical transition from liquid to crystal, for instance, an order or network appears, that of the crystalline structure, and density undergoes a discontinuity. Another classical example is the transition from the paramagnetic to the ferromagnetic in a spin system: when we approach the critical temperature, Tc (Curie point) the global magnetic moment, which serves here as an order parameter, becomes non-null. Phenomenologically, it amounts to considering that the entropic component of free energy, dominant at high temperatures because of the thermal disorder thus generated, becomes secondary before the specifically energetic component due to the interaction between spins, of which the alignment diminishes this energy, all the while being at the origin of the appearance of the global magnetic moment. At the same time, since the effect of a perturbation from the distance r is generally in $\exp(-r/L)$, where L is the length of correlation, then the correlations acquire a scope which increases until it involves the system's total volume. In the case under discussion, this correlation length L diverges as $(T/Tc - 1)^{-\nu}$, where Tc is the critical temperature of transition and ν a critical exponent, often equal to 1/2. Likewise, magnetic susceptibility diverges with its own critical exponent. We may comment on this by saying that at the critical point, a finite cause induces an infinite effect, or that an infinitesimal produces a finite effect.

As we have mentioned earlier, these critical exponents themselves may be calculated by means of the technique of the "renormalization group" (Delamotte, 2004). This technique is made possible because of the infinite length of correlation, which makes the system scale invariant at the critical point. It stems from these characteristics that the "relevant object," both for observation and for the theory, changes at the critical point: from local (the individual disordered spins, say, correlated only to the closest) the concerned scale becomes global (the magnetic moment of the system taken as a whole, because of the global scope of the correlations).

With these simple examples, we have seen at work the notions of critical state, of phase transition, of correlation length (see Bak et al., 1988; Kauffman, 1993, and their numerous examples and references; more will be said below). The scale of observation thus matters in a crucial way with regard to the greatest possible distance of direct causal interactions between the elements of a system.

To conclude, the passing from the local to the global, in physics, necessitates the overstepping of a critical state[2]; this is understood as a mathematical divergence from the length of interaction, therefore as a point where infinitesimal variations create finite changes (or finite variations leading to mathematically finite changes).

From a conceptual point of view, because of the essentially global determination of a biological object with regard to its local components, we will stress the connection with these critical situations, but while considering them here as spatiotemporally extended, and by considering its boundaries (of physico-chemical states) as co-constitutive of the biological entity. From this point of view, the existence and the maintenance of living organisms

---

[2] We are speaking here of a passage which induces qualitative modifications (the critical transition "produces" a new physical object, like a magnet) rather than only a statistical processing at equilibrium (or even far from equilibrium) of a system with many degrees of freedom. This is the case for a gas within which one can nevertheless define "global" magnitudes by means of averages (temperature, pressure, entropy, etc.).

would then be assimilable to the existence and to the maintaining of a situation (or zone) of extended criticality. From the standpoint of complexity, and as we have already evoked previously, the result would be the situation where living matter, regardless of the level at which it is considered, would present an infinite complexity with regard to the physico-chemical. Furthermore, living matter organizes itself into a hierarchy of these levels, according to several orders of infinity, which can be considered as successive processes of "conceptual renormalizations." In Section 6.3, we will attempt to develop this point of view.

6.1.3 Phase transitions in self-organized criticality and "order for free"

The role of critical transitions for the analysis of life phenomena has already been highlighted by other authors. The idea at the center of the approaches originating from the 1988 article by Bak, Tang and Wiesenfeld is that physics proposes several examples of construction of self-organization close to critical states and far from equilibrium, an idea that was already central to Nicolis and Prigogine (1977). Another view regarding self-organized criticality sees it as emerging from chaos: "order for free" to use the terminology of Kauffman (1995) (see also Solé and Goodwin, 2000). In short, to use the point of view expressed in Kauffman (1995), the simple interaction between elements that are not organized *a priori* can be at the origin of the formation of organized structures. It is the network correlating *simple* elements which produces the *emergence* (a keyword in these approaches) of *complex* structures or phenomena. And this, by reaching a critical point: the early and paradigmatic case, analyzed in Bak *et al.* (1988), is provided by the sand piles (dropping sand on a point until the formation of a pile and then of avalanches).

The analyses which we will develop here share with these approaches, especially the one proposed by Kauffman, the role attributed to the emergence of coherent structures during "phase transitions," as a passage from one state, or process, to another. "Life may exist near a phase transition," as Kauffman (1995) explicitly proposes: it appears to be located between order and chaos, at the "edge of chaos," because too much order, or equilibrium, corresponds to death, whereas chaos is the opposite of organization (and living phenomena is very organized). In the cases studied, the notion of the "edge of chaos," a passing point between order and disorder, has a very specific physico-mathematical significance and always refers to a mathematical point in the space of the control parameters (or most often, in the space of *the* control parameter).

What is particularly important in these transitions, which are *critical* in the mentioned mathematical sense, is that the global structure, which constitutes itself (emerges) from the transition, is completely involved in the local activities and vice versa: the magnet, which forms at the critical temperature, at the Curie point, correlates the local spins and the global orientation. In short, the local situation depends on the global one, by a "coherent structure" emerging at the critical parameter. Mathematically, this is expressed by the fact that the correlation length formally tends towards infinity or that infinitesimal variations of the control parameter induce finite changes in the observables; physically, this means that the determination is now global, not local.

The idea of using the physics of criticality, with its "formation of coherent structures" to analyze the emergence of the *structural stability* of living phenomena, is also at the center of our analyses; because over the last twenty years, thie physics of criticality has been developing an appealing theory of the emergence of organized forms in the presence of critical transitions. These structures "constitute themselves all alone" ("order for free"): hence, criticality self-organizes.

Here are a few characteristics of this self-organization, as seen by the authors we have mentioned:

- the existence of many stable states;
- the passing from one to the other of these states may depend on fluctuations (infinitesimal ones);
- the existence of bifurcations, which present themselves at very specific values of the control parameters (critical values); by exceeding these exact values, and only in these cases, coherent structures appear.

In the paradigmatic case of the formation of the heap of sand, when a certain *critical* angle, a mathematical singularity, is reached by the addition of sand grains, avalanches will form, according to a very specific mathematical structuration. One of the limits of this example, at the center of the analyses by Bak, is that self-organization amounts to spontaneously reaching (for free) a critical angle and oscillating around this angle. Moreover, these dynamic sand piles are dissipative-morphogenetic processes, but they are not inherently dissipative: the sand pile stabilizes if the flux of energy/matter stops; particularly, this distinguishes them from the autopoietic and homeorhetic phenomena essential to life. Nor does there seem to be a global coherence structure or a global correlation length: the arrival of a new sand grain produces avalanches *at all scales*; a very interesting phenomenon, but one that is insufficient to develop the views on life we will focus on. As a matter of fact, the activities of an organism are surely not scale invariant: different levels of organization (cells, tissues, organs,...) are understood by different mathematics, if any, as they yield different causal structures, strongly correlated though by integration and regulation.

In other more complex examples, however, it is interesting to note that the formation rules can change over time and do so according to the characteristics of the system: the dynamic is "adaptive" (Solé and Goodwin, 2000).

In general, the physics of criticality enable us to address many cases where interactions matter more than individuals. Now, there is no doubt that biology (and cognition) are above all an issue of interactions, of formations of organized (coherent) structures that are relatively stable. Moreover, critical thresholds spring up everywhere: from genetic mutations, which propagate only from a certain probabilistic critical threshold, to the action potentials of neurons. Likewise, one can analyze many cerebral activities in terms of attractors. Coherent structures then appear beyond the "edge of chaos," in specifically chaotic situations, where these attractors form and disappear dynamically, in an excitable environment. The slightest perturbation can entail the crossing of a critical threshold, typically at the edge of chaos, which triggers a spate of different activities, and the theory of criticality can also provide us with information on this. Is it that the methods of physics suffice to make these aspects of living phenomena intelligible?

We believe they do, in part, because they provide a new and interesting venue for the analyses of the complexity of life and of its "physical singularity." However, the framework remains completely internal to physics and not a single one of the examples in the literature requires different tools than those used for inert structures, maybe with the exception of Kauffman's correlated landscapes, to which we will return. In fact, most of the sufficiently mathematized examples are of a physical nature or represent only the "physical friction" of living phenomena with its interior or with its environment: an essential component of its being in the world, albeit highly incomplete. Let's try to explain this by means of an example. The fractal structure of many organs (lungs, vascular systems,...) is the result of a problem of optimality (maximize the exchange of energy through a surface, in a volume ...) where the physical structure forms in the presence of a critical threshold. This is essential for the growth of living organisms, but the interaction is physical: similar optimal structures can be reproduced with inert materials (beeswax, the fractality present in many physical structures or digital simulations ...). These structures are "geodesics" in the *appropriate phase space* (parameters and observables), as is the case for any physical evolution; one of their main characteristics resides in the punctuality of the critical passages, as the critical values of the parameters are mathematical *points*. All phyllotaxis, with its very interesting and highly developed mathematics, represents this friction between the thrust of life and the physical environment. It always describes, and this is no coincidence, phenomena at the level of organs, which are the locus of the exchanges, of energy in particular, between the living organism and the physical environment (see Bailly and Longo, 2003). Once more, the use of the physics of criticality by Bak, Kauffman, etc., is of great interest. It gives a possible insight into the emergence of life from the inert (Kauffman, 1995). It enriches the analysis of Darwinian evolution, rooted in selection, by a motor that it was lacking: the construction of order from disorder, on the basis of solely physical considerations; let's mention, for example, the networks (metabolic, informational ...) which stabilize themselves beyond a certain threshold, or after the emergence of structurally stable correlations. In fact (and the demonstrations of this are now numerous), the randomness of mutations is insufficient for understanding phylogenesis: the spontaneous formation of order, "order for free," provides a fascinating (and relevant) alternative – an order which is then submitted to selection, of course.

These analyses, however, stop at the "threshold of physics": they bring us to the "edge of chaos," which remains a punctual transition, as everywhere in physics (a very specific value of the control parameter). This punctuality is quite necessary to the physico-mathematical methods widely used in this field: the renormalization group. One of the most telling examples in this regard, percolation (Solé and Goodwin, 2000), is interesting precisely and *especially* for the punctuality of the critical transition; but this punctuality is very general (see Solé and Goodwin, 2000). The modest number of parameters, sometimes a single one, according to which criticality is examined, is also an essential aspect (sometimes said to be positive) of its approaches and of its renormalization methods. The formation of life itself would then be a critical transition, with regards to one or two parameters; and life would situate itself, with its specific phenomena, within a zone that is *close*, but *beyond* the critical threshold.

Now, from our point of view, the physical singularity of the phenomena of life consists, among other things, in the robustness of criticality, which tolerates an *extension* of the critical passage. The punctual "edge of chaos" of physical analyses no longer appears to be sufficient: it must transform itself into *an interval* relative to all relevant parameters (very numerous: time-space, temperature, pressure,...). In a certain sense, the works mentioned attempt to make intelligible the formation of life, from the inert, as a critical passage, but they do not examine as such the lasting phenomenon of life, its "extended criticality." Let's also mention that the conceptual singularity of life also resides in the dynamic of the phase space. We will return to this.

To summarize, the mathematical challenge, with regard to current physico-mathematical theories, consists in the non-punctuality of the structural stability of life (extended criticality relative to numerous control parameters) as well as in the difficulty of establishing a fixed landscape (phase space), within which any process would unfold, following geodesics punctuated by critical transitions. In this regard, it is the phase space itself which changes dynamically (a new organ, a species – unexpected observables and parameters – grow during the course of ontogenesis, of phylogenesis): the dynamic can also be found in the very observables and parameters of the ecosystem, a co-evolutive framework where the emergence of novelty changes the basic situation. Kauffman's notion of "correlated landscapes," with their dynamics, and "adaptive dynamics" (Solé and Goodwin, 2000), are close to the idea we mention in this text; they are quite rich, particularly from the mathematical viewpoint, because the first, for example, specifies a very interesting notion of "correlation" within a phase space normally considered as governed by randomness. They appear however to force a conceptual and mathematical stability, specific to physical theories, since the space of parameters and of observables is given *a priori*. This is insufficient in our opinion to grasp the evolutive dynamics of living phenomena (see below).

On the other hand, within these changing evolutive spaces, what is relatively stable and robust is the lasting or extended criticality of the living object, in contrast to the singularity of criticality in physics: as long as it remains within a range of possibilities (by adapting) the living individual survives along a generic path (a possible, but not necessarily optimal one) within spaces (ecosystems) that are intrinsically changing. Our conceptual outline therefore locates itself at the limit of the physical theories of criticality and attempts to grasp certain aspects of the phenomenality specific to life, in the hope that a unity (or a critical transition) may then establish itself with the concepts and, above all, the mathematics of emergence and of physical self-organization. We finally observe that the notion of extended criticality, as developed here, was first hinted at in Bailly (1991b), whose scientific reference goes back to the early work on *self-organized non-equilibrium systems*, summarized in Nicolis and Prigogine, (1977), more than to the approach by the authors above (see also the end of the next section, where we further stress the difference).

6.2 Life as "Extended Critical Situation"

We have seen that a critical state, in physics, is a singularity within a process: that is, that all along the process, a position, a configuration, a "state," which may be assumed briefly before the global state of that which we are observing changes (radically). A critical state can also be seen as a bifurcation or, in certain cases, as a catastrophe, in the sense of Thom, particularly if this change becomes irreversible. In a sense, it is the opposite of a situation of equilibrium (opposite sign in the mathematical description); and it differs also from the notion of "being far from equilibrium," since this situation, generally, does not imply a "*possible different evolution*" of the system (bifurcations).

We consider life phenomena as being far from equilibrium and continually submitted to perturbations. Thus, by definition, in current physical theories, these systems cannot remain "for long" in a critical state: the development over the course of time, or the intended parameter, forces it beyond that state; this is the pointlike nature of criticality. Also, in physics, criticality is a typical aspect of transition, where minor fluctuations, possibly below the level of observability, can lead to a radically different evolution. The instantaneous nature of criticality has been very well expressed in mathematics by the divergence (towards the infinite, see previous) of functional descriptions, according to specific parameters. Or also, as we have previously discussed, by the maximum of complexity, which also creates instability.

In contrast to physical situations, it very well seems that life, which we may represent, at least for a certain duration, as a stationary state far from equilibrium, evolves within an "extended critical zone," which lasts over time and of which the criticality could be represented as a dense, even continuous, set of critical points in the phase space. This trait would be made possible by the spatial organizational enclosure of an organism, by its finitude in time and by its homeostatic autonomy (integration, regulation). Also, as we have highlighted elsewhere in more detail (Bailly and Longo, 2008), living matter is characterized by the coexistence and the articulation of several types of temporality (cycles for internal clocks, relaxations for the phenomena of stimulus/response, internal or external). Corresponding to it is also a hierarchy entangled with levels of organization of differing natures (biolons and orgons, in our terminology) and the change from one level to another is comparable only at a conceptual level (though a useful comparison, we believe) with that which operates in physics, by means of renormalization for instance, between the local and the global at a given critical point: there exists characteristics intrinsically different amongst similar systems depending on levels and scales. This requires the introduction of concepts beyond those of physics, such as, we have seen, those of *integration* and *regulation*, coupled to that of biological *function* on the one hand, of contingent finality or of anticipation on the other (which at the same time modifies the causality regime of living organisms with regard to that of physics).[3]

Our thesis, therefore, is that the notion of physical criticality could help us to understand a biological situation as a physical singularity "of long duration," "an extended critical situation," in which, particularly, homeorhetic processes maintain a permanent tension between the local and the global. A physical situation which, mathematically, is understood as the locus of a divergence, of a passing from the finite to the infinite, and as the singular point where the relevant object changes (renormalization, see note): the local is integrated into a new object, the global object (the living unit or biolon). Thus, *in physical terms*, the measurement of the objective complexity of the slightest of living units, a cell, a global structure and its components, has an infinite value, in a specific mathematical sense. Likewise, if we consider physical parameters as for measurements (length of

---

[3] An image probably enables us to better illustrate what we mean by that: if we represent the efficient causality of physics as an arrow extending from an initial point to a final point, on a line, the line would be infinite and would thus form a non-compact support; biological causality, which would add itself to physical causality, could then be represented by a similar arrow but on a closed line, this time, forming a compact support, representing the effects of retroaction and of finalization (possibly also related to the effects of enclosure – and of autonomy – which have been evoked). Passing to another level of analysis, we could consider that this type of causal manifestation takes place on the internal fibers constituting additional compactified dimensions. The two other causal structures are of course compatible and "simultaneous," because living phenomena are immersed within physical fields.

correlation and its relative effects), the living element is in a progressive critical situation, a permanent passing between the local and global. It is thus that a singular dynamic unit, that of living matter, from the cell, is infinitely more complex than any physical process, which may behave in a critical fashion only in exceptional cases, of short duration, as mathematical singularities.

To summarize, a unit of living matter, a biolon according to our terminology, critically unstable, is preserved in its extended situation, far from equilibrium, by homeostasis, or better, by homeorhesis. Or else, the dynamic integration and the regulation of its components (orgons, with their components, biolons, with their orgons ...), their "ago-antagonistic" relationships (Bernard-Weil, 2002b) within themselves and their environment, sustain them within an improbable physical state. Autopoiesis (Varela, 1989; Bourgine and Stewart, 2004), constitutes another way of expressing this auto-constitutive dynamic (see Section 6.3 below). A mathematical organization, which may be related to autopoiesis could refer to several coupled endomorphisms, in mathematical terms; their "organizational enclosure" may then correspond to the limits of structurally stable attractors; the insides and the membranes may be understood respectively as the basins and the edges of attractors (or, to be more precise, as the physical expression of the proper attractors, which are given in the phase space).

From the moment that integration or regulation no longer works or exceeds the limits of the pathologically tolerable situation (limit of functional plasticity), everything collapses: entropy suddenly grows, disorder represents death. Running upon a tightrope is a good representation of the progressive situation of a biolon: when control, as regulation and integration, decreases to a certain level (the critical boundaries of the extended critical situation, as acceptable limits of pathology) death terminates this contingent life, by a final and irreversible transition state. Of course, at the boundaries of the extended critical situation, phase transitions, changes in correlation length, passage through singularities ... continually occur, but, within the considered limits, they are confronted by a regulating activity. In fact, they form an essential part of it: all biochemical thresholds, which contribute to a biolon's internal exchanges, may be perceived as elementary components of global homeorhesis. More globally, life itself can be seen as an "extended physical singularity."

Physical paradigms have helped us to formulate this notion, which is not of a physical nature. Monism, intrinsic to modern science, refers to matter, and not to methodology; in fact, we face different phenomenalities and attempt to make them intelligible by organizing them by means of various notions and conceptual structures (and, if possible, mathematical ones). A synthesis is far from obvious; it must be constructed by means of a new (and mathematical) conceptual unity, as a long-term objective.

As we have already discussed, the formation of such an extended critical zone, associated with the conditions of possibility for life, could proceed from the convergence of two processes whose origins are opposed: on one side (first boundary of the zone), following Nicolis and Prigogine (1977 and 1989), starting from equilibrium, a series of critical bifurcations, which could be generated by fluctuations while moving away from equilibrium and, on the other side (second boundary of the zone), the self-organized criticality stemming from the stabilization of originally chaotic states (see Kauffman, 1993 and 1995[4]). In short, we consider that both approaches, ranging say from Prigogine to Kauffman, are extremely interesting, yet, we believe that each of them describes only "one side" (one "conceptual boundary") of the phenomenal situation of life: in our approach, they coexist and bound the extended criticality of living systems, in the space of parameters. Moreover, the possibility (and constraint) of the coexistence of these two different boundaries would be precisely associated with:

- the formation of levels of organization interacting among themselves,
- the organizational enclosure of autopoiesis,
- the constitution of a homeostatic autonomy capable of maintaining itself.[5]

---

[4] As for the approach in Bak et al. (1988), observe that extended critical situation can only exist and maintain itself far from thermodynamic equilibrium and in the active presence of exchanges of matter, energy, and information with the environment. In this, it comes close to the conditions within which dissipative structures constitute themselves while at the same time distinguishing itself from self-organized criticality (exemplified, namely, by the behavior of sand piles) such as has been specifically studied in Bak et al. (1988). As we mentioned above, the latter are dissipative-morphogenetic processes, but they are not inherently dissipative (the sand pile stabilizes if the flux of energy/matter stops) nor yield coherence structures.

[5] We know that in physics it is the boundary conditions which generate (for vibrating strings, for instance) the discrete stationary states that are solutions to propagation equations. Maybe we can find a metaphor for the constitution of these levels of biological organization related to both the constraints of organizational enclosure as well as to those of having to evolve within an extended critical situation bounded by the characteristics, which we have just evoked (sequences of bifurcations, on one side, chaos, on the other).

6.2.1 *Extended critical situations: General approaches*

What can it mean, for a system, to have to evolve within an "extended critical zone"?

First, it must be to find itself at all times far from equilibrium because the maintaining of its organization requires intensive exchanges of energy allowing it to maintain an "abnormally" low entropy with regard to the situation of equilibrium.

Next, and correlatively, to present itself with an internal organization, that can extend to the constitution of levels of organization, corresponding to the constraints of dissipation (dissipative structures) within the context of these exchanges.

Third, to form a "whole" within time and space (at least locally and momentarily), inasmuch as its internal correlation lengths are of the size of the system itself (critical situation) and its characteristic times, of more or less exhaustive running through the attractor corresponding to its situation, are bounded, thus defining the scale of its existence (namely by the maintaining of its internal organization). In this way, therefore, is manifested a first form of spatiotemporal autonomy, coupled to this fundamental heteronomy, which constitutes the previously highlighted necessity of exchanges with what lies beyond. In this, also, the system presents itself as an "extended" singularity in the usual physico-chemical landscape (possibly calling for a theory of generalized catastrophes, an extension of the theory of elementary catastrophes).

Does evolving in such a zone impose the existence of an organizational enclosure and the establishment of a boundary? An argument in this direction would be related to the necessity of limiting leakages and of regulating the exchanges of energy; another, within the framework of an autopoiesis, would be related to the necessity of implementing boundary conditions enabling us the very existence of such a system; a third could be found in the fact that the existence of an extended critical zone would only be possible upon a topology enabling to distinguish an inside from an outside (compacity argument?). These arguments are somewhat circular inasmuch as one supposes the existence of the system in order to define the conditions for its existence. This must signify that the hypothesis, which consists of associating living phenomena to a zone of extended criticality, if it enables us to interpret the permanence and the functioning of organisms, still remains insufficient to describe or to explain the origin. What appears, at this stage, namely by the fact that this hypothesis says nothing about learning capacities – even if in certain regards the quest for structural stability can somewhat resemble them – and especially nothing concerning anticipatory capacities (or contingent finality) – even if the spontaneous implementation of a timescale characterizing its own limits of existence may be considered as a prefiguration of protension (a time scale may correspond to an "extended present," thus to the time of "protension," which assumes learning, see Varela, 1999). That is, life cycles and rhythms may relate to Varela's Husserlian "extended or specious present" or to Rosen's anticipatory capacities, an issue to be further investigated.

In what concerns the objectivity of these constructions, it is first an issue of distinguishing, as usual, within living phenomena themselves aspects of structure (anatomy, for example) and dynamic aspects (physiological functions, for instance). Then, also of having *particular ways of viewing* living phenomena in general, either structurally (a-temporally) or dynamically as above, associated also with genesis and duration (see, for example, the theory of viability (Aubin, 1991)).

Different approaches, no longer solely conceptual but also more or less mathematized have for a while now been distributed according to this partitioning, without their articulation having been solidly ensured, to our knowledge. Among the most recent, let's mention on the one hand the applications of the theory of singularities (elementary catastrophes) by Thom and on the other hand the application of the thermodynamic theory of bifurcations by Nicolis and Prigogine. One of the possible contributions should consist, for the structural aspect, in the recourse to a mathematization within the framework of category theory (also a-temporal or structural) and, for the dynamic aspect, to develop a theorization of the extended criticality that we have just evoked by introducing the couplings and scales of relevant temporalities. A project on the articulation of the two approaches could introduce the relevance of temporalities in the "categories" (structural) aspect and a stable genericity (see below), that is some structural stability, in the "extended criticality" approach.

But let's, nevertheless, try right now to be more specific and more precise in our attempts to characterize what we mean by an "extended critical situation."

6.2.2 The extended critical situation: A few precisions and complements

6.2.2.1 Spatial aspect: Concerning correlation lengths

The "correlation lengths" for living matter seem to be closely related to the transportation properties of the various molecules and substances which actually enable them "to pass information" from one locus to another, or to modify the biological behavior in one location because of a stimulus originating in another. Now, depending on the size of the biolon, there seems to be two types of transportation processes likely to occur. For larger organisms, that would be of the "propagative" type (velocity v_p) and the typical correlation length will be $L_p = v_p\tau$, where τ represents the characteristic amount of time. For small organisms (cells, for instance) we would have a "diffusive" type (diffusion coefficient D) and the typical correlation length would be $L_d = (D\tau)^{1/2}$. (Neurons, of course, present the notable exception of "diffusing", also, action potentials.)

We notice the difference in dependency according to time: linear in one case and in a $1/2$ exponent in the other.

Two complementary remarks:

- The size of the organism is also at play in the nature and the implementation of the selected structures to ensure the mode of transportation; for example, in the case of the respiratory function (transportation of oxygen) in small organisms (insects, for instance) the transportation takes place through tracheas (or pores), a multitude of little cylinders where the air is diffused in order to supply the cells with oxygen; in larger organisms (fish, mammals) the transportation and exchanges take place by means of gills or lungs, "centralized" anatomical structures presenting fractal geometries enabling us to reconcile the hardly compatible constraints we have already mentioned (efficiency, stericity, homogeneity); the transportation, in this latter case is also much more of a type which we can qualify as "propagative" (even if diffusion does play a part for bronchioles, for example).
- These considerations are mainly valid for different *structural* aspects relating to identical functions. The *functional* aspect, on the other hand, is very generally governed by common scaling laws, which we have already discussed (metabolism, which corresponds to oxygen consumption, the variegated rhythmicities, the periods of relaxation, ...). It thus appears that the modes of transportation associated with iden-

tical functions can differ and can correspond to different anatomical structures.
- Finally, having taken account of these remarks, if the characteristic times τ generally scale as $W_f^{1/4}$, where W_f corresponds to the average size (or mass) of the adult organism, one must expect the correlation lengths to scale differently according to the modes of transportation: respectively L_p in $W_f^{1/4}$ and L_d in $W_f^{1/8}$.

6.2.2.2 Temporal aspect: Return to the specificity of the temporalities of living matter

We outline here an idea developed elsewhere concerning biological temporalities (Bailly and Longo, 2008). The causal intrications of the various levels of organization of living matter have led us to think that the extended critical situation – corresponding to the self-referring and individuated character of the organism – presents a topological temporality of the **RxS₁** type, where **S₁** is a circle. Thus, **RxS₁** is the continuous time of the real straight line times a compactified time-line, a circle, **S₁** – to be naively understood as measured by an angle, as on a clock. This is the internal time, where one may represent the internal clocks, while the organism's externality (and the way in which this externality reacts with the organism) preserves its usual temporal topology **R**. The idea is to provide a "reference system" (the "axes for representation, in the Cartesian sense) accommodating physical time *and* biological rhythms (see Bailly et al., 2010, for more and references). Thus, from the standpoint of time now, the extended critical situation would be characterized by this bi-dimensionality, being coupled, from the spatial standpoint, with quasi-infinite correlation lengths (at the scale of the whole organism) and the organizational enclosure would conjugate these specific spatiotemporal aspects, which would manifest as autopoiesis. The extended critical situation would therefore delimitate, from the standpoint of space-time (to which we will add all other relevant parameters such as temperature, pressure, the quantity of nutrients or of oxygen, ... see next), a sphere of a spatial radius L for instance, where the correlation lengths of the interactions would be of the order of L, and the "temporal radius", θ, where varied cycles and rhythms may be accomodated. This gives a two dimensional time $\tau_v \times \theta$, that is "lifespan" which multiplies a "radius".

Without changing the essence of the question, we can, nevertheless, present a slightly different way of seeing it: for a living organism, the extended critical situation would occupy a volume within an n-dimensioned "space." Among these n dimensions, we could distinguish three spatial dimensions as such ($\mathbf{R}^3$ topology) and two temporal dimensions ($\mathbf{R}\mathbf{x}\mathbf{S_1}$ topology) of which the compactified dimension takes a non-null radius outside of this volume, the remaining n-5 volumes corresponding to the compatible values of the vital parameters (temperatures between T_1 and T_2, metabolisms between R_1 and R_2, etc.). The metrics of the volume space maximally correspond to the life-spans (for $\mathbf{R}$) and to the pure maximal numbers (maximal endogenous frequencies) for $\mathbf{S_1}$ (for more on this, we refer the reader to Bailly and Longo (2008).

One will notice that the endogenous rhythmicities and cyclicities are not rhythms and cycles (whose dimension would be the inverse of time), but *iterations* of which the total number is fixed regardless of the empirical life-span, at least for a metazoan.

6.2.3 *More on the relations to autopoiesis*

Our conceptual frame may also be understood as a way to further enrich and specify the notion of autopoiesis (Varela, 1989). Autopoiesis doesn't imply extended criticality, nor the other way round, but the two notions together may join in an effort to focus on key phenomenal properties of living systems. Recall that an autopoietic structure is

"a network of processes of production (transformation and destruction) of components which: (i) through their interactions and transformations continuously regenerate and realize the network of processes (relations) that produced them; and (ii) constitute it (the structure), as a concrete unity in space in which they (the components) exist by specifying the topological domain of its realization as such a network".

(Varela, 1998)

Thus, autopoiesis may be seen as a way by which the living generates and preserves extended criticality, far from equilibrium, yet in a functionally coherent structure. The *qualitatively* constant internal environment (in relation to the external environment) is part of the structural stability or stability of the *qualitative* organization as captured by autopoiesis and departs from cybernetic feedback or the like. Autopoiesis contrasts the increase of entropy due to the metabolic/thermodynamic irreversible processes, by improving/maintaining the organization (a growth/preservation of organization decreases the corresponding entropy). And this is a crucial "physical singularity" of living systems: the capability to lower entropy, along an irreversible process, by (re-)constructing organization. A mathematical approach to these ideas is in Bailly and Longo (2009).

As already observed, both autopoiesis and extended criticality require an "operational closure," that is a delimited space-time for the evolution of the process. Extended criticality proposes it by the (limits of the) coherence structure proper to critical states (due to correlation lengths). The peculiar dynamics of autopoiesis may be understood in the terms of the local instability of criticality, which we also described as a dense set of critical *points* in the phase space associated with a global structural stability, and limited in space and time: an autopoietic system continually (and *locally*) moves from an unstable state to another, while preserving *global* coherence (or structural stability). This feature may help us to understand its ability to explore various local surroundings (ecosystems) and to adapt to new (internal or external) constraints, which appears as plasticity properties. We will go back to autopoiesis in the Intermezzo below.

6.2.4 *Summary of the characteristics of the extended critical situation*

To summarize certain aspects encountered and discussed here, we could characterize the extended critical situation by means of the following (non-exhaustive) traits:

- a spatial volume enclosed within a semi-permeable membrane
- correlation lengths of the order of magnitude of the greatest length of this volume,
- a temporal bi-dimensionality, one classical, bounded between conception and death, associated with the bio-physicochemical evolution of the organism in relation to an environment (behavior which can be of the relaxation or of the cyclical types), and the other compactified, associated with the endogenous physiological rhythms of the organisms, manifested by dimensionless numerical quantities,
- a metabolic activity far from equilibrium and irreversible, involving exchanges of energy, of matter, of entropy with the outside as well as the production of entropy,
- a confinement within a non-null volume of a space of parameters (temperature, etc.) of n dimensions,
- an anatomico-functional structuration into levels of organizations, which are autonomous, but coupled amongst themselves, likely to be

distinguished by the existence of fractal geometries (membranal or arborescent); the fractal geometries can be considered as the trace (or model) of effective passage to the infinite limit of an intensive magnitude of the system. These levels of organization alternate biolons and orgons (the latter being essentially the locus of manifestation for fractal geometries). The lengths of correlation manifest both *within* and *between* these levels.

6.3 Integration, Regulation, and Causal Regimes

A few additional facts may also relate physical and biological phenomena, while also highlighting their differences. Causal relationships are local in physics; they may be global only in the sense of a field connecting physical entities by the propagation of local interactions. Thus, in this approach, the global structure, with its correlation length, is only obtained by means of the transitivity of local interactions. This way, in each of the physical cases (local/global), mathematics, by fixing a phase space isolates a unique level of causality. In biology, conversely, the local causality may radically differ from global correlations, and yet cannot be isolated from the latter: integration and regulation, typically, causally affect local interactions (local biochemical exchanges can be regulated by hormonal cascades or by neural signals of a completely different nature). In other words, the global causal regime may differ from the local regimes, even if they permanently interfere with one another: thus, the unity of a biolon is given by a regulatory-integrative activity, of a different physical/biochemical nature with regard to the exchanges between and within orgons, althought these exchanges are regulated (and affected) by the global. Notice that it is the global regime that essentially contributes to homeorhesis (autopoiesis, ago-antagonistic couplings) as the maintaining of an extended critical situation; its fluctuations, within the boundaries of criticality, correspond to the pathologies that a biolon may live with, as well as to some forms of local death (cellular apoptosis within an orgon). Naturally, fluctuations exist in physics (and can be "tolerated"), but the underlying physical entities do not change over the course of the fluctuations. In this sense, in physical theories, there is nothing that resembles phenomena such as pathologies or local apoptosis; the latter's mathematics still remains to be invented, as much as their pre-formal conceptualization.

To summarize, *integration* is the presence (upwards-causally) of the local within the global structure, whereas *regulation* is the global structure causally affecting (downwards) local structures. It is this local/global interaction which organizes the "internal conduct" and which maintains the extended critical situation at least at the phenomenal level (see also Rosen, 1991; Stewart, 2002). But besides having this role of stabilization, these interactions also participate in what we will later define as a (possible) theoretical indetermination of the evolution of living matter (both throug phylogenesis and ontogenesis), because they interconnect different levels of organization, each governed by its own regime of physical causality: we will then also attribute indetermination to the "resonances" between these various causal structures.

Despite the physical singularity of living phenomena, let's attempt once more to define our understanding in familiar physical terms. Regulation may play the role of initial boundary conditions for the global behavior of the solutions of systems of equations (differential or of finite differences), that is, dynamic systems described by these equations. Integration may also be understood, by rough analogy, as the correlations of variables conferring unity to a given system of equations, or, also, in its role of organizing singularities in their solutions. It may also be understood, likewise, as the analytical extension of a locally defined solution. Once more, we are only attempting to approach, by means of physico-mathematical concepts a phenomenology which, till now, extends far beyond these descriptions, in the mathematical sense of an infinite complexity, as hinted above (the complexity of a singularity).

Intermezzo: On contingent finality and entropy

Inter 1: Finalism

In our view, biological causality needs to integrate some notion of finalism, in a theoretically autonomous approach to living systems. When and if reduction to physical theories will be fully accomplished, then (a suitable) physical causality, possibly time-oriented, will surely "explain" in more appropriate terms, the various circularities of life, including the causal ones. But this may require a change or an enrichment of current physical theories as much as quantum mechanics and relativity are ongoing deep revisions towards unification (of space and time – non-commutative geometry – or of the very objects – string theories). In the meanwhile, for our theoretical purposes, we called "contingent" the conceptually needed finalism of life as:

- it doesn't need to be there (it is not necessary: life or a specific living being may not exist nor last);
- it is *local* not global, in a sense we will try to explain next.

The natural way to see finalism in life is given by anticipation or protension, in phenomenological terms: an anticipated event *causes* or influences an ongoing process. More deeply or at the simplest level, a metabolic cycle *tends towards* its own preservation, when it takes place in a living organism – a basic form of protension. Again, this may be seen as part of the definition of life: life exists exactly because it tends towards preserving itself – otherwise it would simply not be there – this is its contingent finality.

In this sense, autopoiesis may be seen as a contingently finalistic process: the interacting network of processes *tends towards* the productions of the components that produce the processes. Otherwise, since it is far from equilibrium, as soon as the "effort" of maintaining itself in a permanent flow of energy stops, it would collapse in that state of equilibrium, which is called death. Autopoiesis is a way of describing the maintaining and/or improvement of metabolism, which is the (minimal and) contingent finality of life.

In contrast to old-fashioned Leibnizian finalism, we are aware that causality in modern physics is largely analyzed in terms of "symmetries" and symmetry breaking (see Chapter 4). In particular, symmetries correspond to conservation properties, thus they realize the geodesic principle in various contexts. In the theoretical analysis of life, it seems that the addition of a further conservation principle is required (or useful): "preserving autopoiesis" (that is life itself). By its nature, this may seem finalistic and may break physical symmetries. In particular, the geodesic principle doesn't seem to govern the directions of evolution in so far as these are only *possible* (generic) paths: the phylogenetic (and ontogenetic) drift goes along all possible (*compatible*) directions, it does not follow a (unique) geodesic. One more reason to consider insufficient a causal analysis based only on symmetries and geodesics is that the finalistic preservation of autopoiesis (and thus metabolism) "forces" species and living objects to go along all possible evolutive paths.

Inter 2: Entropy

Let's now further argue about the *local* nature of this new conservation principle that life implements, in reference to entropy. In our view, the relevant property of autopoiesis is that not only dors it produce networks that produce components that produce those networks, but also that it contrasts the inevitable growth of entropy related to the intended and irreversible flow of matter and energy. As a matter of fact, there are two kinds of entropy that participate in the process. On one side, there is the thermodynamic production of entropy related to irreversible energy consumption, an entropy which, by this, continually grows (Nicolis, Prigogine, 1992). On the other, autopoiesis decreases entropy by increasing or maintaining organisation. Organization, thus negative entropy, increases at least during embryogenesis; later, it is maintained, though aging may slow the (re-)organization process. Death supervenes when the (re-)organization process cannot oppose anymore or not sufficiently, by decreasing entropy, the increase of the other kinds of entropy.

In short, on one side, entropy grows as a consequence of irreversible physical processes. On the other, entropy decreases by a permanent (re-) organization of the living material (formation of tissues and organs first, their maintenance and permanent renewal later). Now, these contrasting productions of entropic processes manifests themselves *locally*: at each instant autopoiesis produces enough organization so as to prevail, in the game of entropies. More or enough organization is produced so as to continue the game, unless pathologies modify the dynamics or lead to death. However, this very game is an irreversible process, it thus produces its own entropy. And death is the inevitable end of it. Yet, and this really matters to us now, there is no need to propose a *global* finalism to understand autopoiesis: as we said, at each instant autopoiesis goes along the contrasting line of entropies, and a *local* analysis of its evolution makes it understandable.

This understanding is not so different from the turn that separated physics from finalism, at the beginning of the XIXth century. Till then, light was seen to go along the shortest path, *since* it was following the best way to go *towards* an objective. Similarly, inertial movement was often described in finalistic terms: the choice of a geodesic or of the optimal trajectory in order to go somewhere. The situation radically changed through an understanding of geodesics as a (mathematical) integral (a sum) of local properties. To put it briefly, a straight line is defined locally by points where the *tangent is constant*. Similarly, conservation of energy or of other quantities is a local, point by point, property (Galilean inertia is a local): the global path is obtained as an integral (a sum) of the local values. Thus, finalism in physics was put aside as inertia, typically, or "going straight" or along a geodesic is the result of a (*previous*) or present history of the phase space and nothing else.

We are not (yet?) able to perform this in the analysis of life and we need to integrate in our theoretical proposals the contingent (local) *propensity* to contrast the growth of thermodynamical entropy (due to the irreversible flow of energy) by the decreasing entropy of a permanent (re-)organization of living structures. Yet, we see this as a local form of finalism, which realizes itself instant by instant and does not need to assume any global, pre-established finality.[6]

6.4 Phase Spaces and Their Trajectories

Biological situations would nevertheless present an additional trait, rather essential in our view: as we have already emphasized, this spatiotemporal extension not only concerns the behavior of "trajectories" within a given phase space, but also reveals itself through a modification, through living phenomena, from this phase space (ecosystem, in a very broad sense) itself. These modifications would concern the interaction with the environment and would be induced by mutations, the appearance or disappearance of certain organs, even of populations, etc. The situation would be rather comparable to the existence (metaphorically speaking) of a sort of set of phase spaces (the one corresponding to the union of all possibilities for a given organism), characterized only by a few great structural invariants (we are thinking here, namely, of the absolute numbers, which transversally

---

[6] A paradigmatic case of global finalism is the theory of programming in genetics (see Longo and Tendero, 2007): if the genome were a program, in the sense of computer science, each cell of a metazoan would contain a well written code for *all* its *phenotypic aims* in life. Programming is the most finalistic activity that man ever invented: a program is *written for* a possibly remote aim. In contrast to this view, we understand genes as traces of a phylogenetic history, which intervene in the autopoietic process by statistical, stereospecific interactions with macro-molecules. By their rigid structure they contribute to set the limits of extended criticality. In particular, they induce the primary structure of proteins as well as, indirectly and by epigenetic phenomena, the proteins' shape – secondary and tertiary structure. In short, genes force the approximate iteration of an history; thus, they contribute to the species' structural stability, in conjunction with the relative co-constituted stability of the ecosystem, by "forming" the molecular material of autopoiesis. Yet, by genetic drift and degeneracy, in the sense of Edelman (see Edelman and Tononi, 2000), they also contribute to variability.

characterize many expressions of the living phenomena, such as metabolic rhythms, for instance). These great spaces of invariants would be themselves partitioned into phase (sub-)spaces (those corresponding to the effective coming into being of these possibilities), the trajectories generally taking place within these sub-spaces along strange attractors. Yet, they may be likely, under certain perturbations (genetic mutations, brutal change of physico-chemical or biological environment), to pass from one sub-space to another, thanks to the properties of these attractors (see 6.6.1).

However, this set of phase spaces (or of *evolution*) would be far from being predetermined. More specifically, it is at this level, firstly, that one should locate a sort of biological indetermination: that is, at the level of the passage from one phase space to another (ecosystem), at a given moment, to that of the following moment; and this passage would "contain" or would express the biological trajectory (phylogenetic, ontogenetic). To explain this, let's briefly return to the analogies/differences with physics.

In classical physics, *a phase space is given* within which form the specific trajectories, actually the geodesics of the relevant objects (see 6.6, further down); in some cases (non-linear, typically) the formal global determination of these evolutions (the existence of equations determining the dynamics) does not necessarily imply the predictability of the trajectories (but the system remains deterministic). On the other hand, in quantum physics, *once the phase space is given* (the quantum states within a Hilbert space), the dynamics does not go along proper space-time trajectories in the classical sense, and it is an intrinsic indetermination which governs the theoretical intelligibility of the phenomena; particularly, the causal relationships are replaced, within the given phase space, by correlations of probability. This conceptual and mathematical framework, with its indetermination and its probabilistic analyses, is at the center of quantum mechanic's theoretical originality (and of its conceptual difficulty).

We believe instead that, in biology, the theoretical and conceptual difficulty resides principally in the impossibility to give, upwards (or *a priori*), a formal global determination and a *phase space* (or a space of evolutions) able to enframe the phenomena for a sufficient amount of time; that is, to describe a global mathematical determination (a set of equations, typically) within a phase space that is *a priori* and *given once and for all*, or at least for very long. In fact, if there exists a relative structural stability, it concerns the objects and some of their components, more than their framework of living, including their ecosystems or spaces of relevant observables and variables.

To put it in other words, in physics, the mathematical notion of a "state-determined dynamical system" (typically expressed as a set of differential equations) does sterling service. Its operational effectiveness stems in large part from the fact that the dynamical laws are fixed, i.e. they are "laws of nature" or "conceptual construction principles" that are given *a priori*, within the theoretical proposal, and it would be meta-physical to ask "why" they have this or that particular form. In addition, for a given system, the list of variables and the boundary conditions are also givens (at worst, in quantum physics, in the presence of the "creation" of a new particle, one gives a Fock space in order to accommodate it, *a priori*, instead of a Hilbert space). The great difficulty is that, in biology, variables, observables, boundary conditions, and even the dynamical laws themselves are actually constituted by the operation of the living organism itself. And this difficulty is multiplied yet again when we wish to take properly into account the interactions between an organism and its environment: because the environment is also constituted by the organism, typically by its "ecological niche." Thus, there is through and through a co-constitution of the very frame of intelligibility. In conclusion, the "bifurcations" take place as much at the level of the selection between possible determinations, as within possible and different determinations.

Let's return to this delicate point, as it happens, in terms of trajectories. In mathematics, the critical points on a curve (maxima and minima, for example), may be said to be *specific* in that they are in most cases much less numerous (a denumerable discrete) than the continuous to which all the other points belong, which can then be said to be *generic*.

In physics, in this regard, within the given phase space, the set of "conceivable" trajectories is generic, while the effective trajectory, defined by the geodesic principle, is specific (critical, stable, meaning minimal for the Lagrangian action, or in the particular case of optics, minimal for the optical path). In other words, effective physical phenomenality is specific and is enclosed within a relevant phase space (a great part of the physicist's job is actually characterizing this space). It is doubtlessly this which confers to physical theory a great mathematical force as well as a possibility, by means of abstraction, to characterize the physical objects using very general properties, despite the singularity of each specific experience of which the conditions are not always exactly reproducible: the trajectory of any object will be specific and its analysis is related to the phase space (abstract and general).

In contrast, it appears that for biology, the cells of an organism, the organisms of a species, the species of an environment (that is, the biolons), are concerned with "generic trajectories": all those which remain *compatible* with their perduration, even their transformations (mutations). It would be the falling back upon the specificity of this generic which would cause it to lose its biological character each time a physical reduction is attempted.

From then on, it is possible to understand the *extended critical situation* as the expression of this genericity in contrast with the localized *critical transition* of physics (within the space of parameters and within time, due to their incessant fluctuations). To put it in other words, there would be a sort of duality between physics and biology: specificity of the trajectories and "locality" of the criticality for physics, genericity of trajectories and extension of criticality for biology. At the same time that would enable us to "naturally" relate the biological to the considerable variability of which it is the locus given that this generic compatibility would allow for the existence of a great number of possible "trajectories." Invariants should no longer then be defined within a given phase space but over the set of phase spaces that are compatible with this genericity, each specificity being able to modify the local structure of this set without, however, disappearing. This is maybe what would explain the great invariants (approximated), which we have mentioned, almost always concerning considerable sets of organisms (mammals, oxygen metabolisms, the animal kingdom itself, etc.), because they would constitute some of the rare invariants with regard to this global plasticity. These invariants endure, within the variability of living phenomena, and contribute to its stabilization; moreover, this biological variability differs from physical variability, because the latter does not normally contribute to stability (at least in non-linear cases) and finds itself within a framework of determination, while the former may go beyond it.

To conclude, we believe that, for biology, it is the collection itself of relevant objects and of parameters (phase space) *which will follow the current situation*, which is indeterminate and that this indetermination should be intrinsic to the theory, as the co-constitution of living individual/species (biolon) with and within an ecosystem is a major theoretical challenge in biology. Within this changing frame, the trajectories themselves (a notion that is preserved although it becomes problematic also in quantum physics), differ from those of classical physics because they are not specific (geodesics within given spaces), but generic (possibilities within changing spaces). In a sense, there would therefore be two levels of inter-correlated indetermination: that of the passing into the new ecosystem and that of the "choice" of trajectory for each biolon, among all compatible trajectories. Over the course of phylogenesis, typically, the formation of a new species, which participates in the ecosystem, even the emergence of a new organ, modifies the evolutionary space, even the space of ontogenesis, and modifies the set of possible trajectories. Once again, the genericity of the trajectory ought not to be analyzed only within the same phase space, but also in terms of passage to a new possible space.

For this reason, in our opinion, the analysis of the dynamics of living phenomena cannot be reduced to the terms of classical deterministic unpredictability (with its geodesics, of which the choice is more or less sensitive to boundary conditions), but to those of an intrinsic indetermination of the evolution of living organisms, to be understood in terms of the genericity of the trajectories and of changes of phase spaces. As a result, similarly as classical dynamics and quantum mechanics propose two causally different notions of randomness (a dice and spin-up/spin-down are random in a very different sense), so our approach to biological phenomena suggests a third form of randomness: the indetermination locates itself at the level of the very space of phases or evolutions and coexists with a relative structural stability of individual biolons.

The extension of critical situations, together with the intrication of the levels of organization and their effects of reciprocal "resonance," proposes an intelligibility of the physical *type*, inasmuch as physics succeeds in providing us with adequate metaphors. As quantum physics has succeeded with regard to classical mechanics, it would be necessary to provide ourselves with autonomous concepts and, if ever possible, mathematical structures, in order to better grasp the fields of living phenomena and their dynamics.[7] *one* phase space. It is there, we believe, that we must shift from a determined trajectory to an intrinsic indetermination, comparable to the possible and indeterminate paths of quantum physics, but concerning the phase space itself. Moreover, as we have stressed many times, within each space, it is the generic trajectories that play an important role, not just the geodesics.

---

[7]The theory of viability (see Aubin, 1991) proposes a mathematical analysis which is close, but different, to the approach which we propose: we replace the evolution equation ($dx/dt = \ldots$) membership within a set of possible evolutions ($dx/dt \in \ldots$). However, the function $x(t)$ and the set, which determines this passage, are given beforehand and contain, according to our interpretation, the list of all possible future spaces, with their geodesics; $x(t)$, particularly, represents a "trajectory" of phase spaces, instead of a trajectory within

6.5 Another View on Stability and Variability

We have mentioned earlier the stabilizing role of regulation and integration for the intertwining and the intrication of levels of organization. Our analysis, in terms of the genericity of the trajectories and dynamics of phase spaces should contribute to making intelligible the fact that the phenomenal life is characterized by the preservation of a few key invariants not only due to stability, but also to variability. For this reason, we have considered the intertwining and the interactions of organization levels on the one hand as one of the components of stabilization, and on the other as forming part of biological indetermination (as resonances destabilizing each level of organization). Following variability, typically, biolons all differ from one another: the individuals of a species are not and must not be identical. Even cells have a certain form of "individuality." Indeed, variability is at the center of evolution and, in conjunction with individuation, it is also at the basis of ontogenesis. Clearly, here lies another singularity of living phenomena in reference to the physical descriptions, where individual entities of the "same type" are all identical and their description can be, conceptually as well as mathematically, perfectly stable (the specificity of physical trajectories, which we have addressed).

6.5.1 *Biolons as attractors and individual trajectories*

Let's now try to understand this approach, described by indeterministic dynamics within globally stable frameworks, and this stability/instability dialectic, in different terms, which are, nevertheless, compatible with the former.

On the one hand, a species presents a global structural stability, whereas its members may display different variations in their nature and behavior. Likewise, in its own time scale, an individual is (relatively) stable, while its cells evolve and die, each according to a different path. In one or the other case, the global structure is essentially (but not entirely) preserved while local variations occur. If we take a physical analogy, the greater biolon (a species, a metazoan) can be described by means of the geometry of an *attractor*, of which the global dynamic is essentially stable as long as it remains within the same phase space. The relative instability of individual trajectories within the attractor, those of the smaller biolons in it (the cells in a metazoan, the individuals in a species), contributes to variability of the global structure. Yet, the local biolons, as part of orgons, are stabilized by the function (of the orgon they may belong to – a population in the case of a species). Thus, the functionality contributes to the local and global stability, by maintaining the function. While their minor instabilities constitute adaptability, by (minor) changes in the individual trajectories (or, say, variations in the fractal structure of lungs yield changes of survival in changing environments).

We can understand the change of phase spaces as the toppling of the attractor into another space, with the characteristics of indetermination, which we have discussed. The difficulty would then consist in the search for the correct classes of universality, or of the great biological invariants, such as, for instance, absolute numbers (the clocks of living matter, etc.), which concern the species, even entire phyla, and which would allow us to speak of a toppling of the "same" attractor.

This dynamic stability is as compatible with the variations and even the instability of the individual trajectories, which may fall within the same attractor (the *smaller biolons*, included within the largest – a metazoan, a cell, respectively). In this sense, a biolon of the intermediate level (a metazoan) constitutes a trajectory for a species, the latter being considered as an attractor; at the same time, this biolon behaves as a stabilizing attractor for its own component, its cells.

As we have observed, in physics, the dynamic of the trajectory is given by the geodesic within a predefined phase space; thus, it is selected according to the highest stability and generally in a unique way. In other words, it is specific and *optimal* in the sense given by the metric space, a measure. In contrast, individual biolons are, in our opinion, represented by different and generic simultaneous paths, and simply submitted to compatibility constraints with conditions at the boundaries (which, in turn, do not have to be stable and can be modified by continuous processes). This approach would enable us to simultaneously capture the properties associated with:

- the interaction of relative global stability and individual variability, as attractor *vs* trajectory;
- compatibility instead of "optimality" of geodesics, as recalled above, given that trajectories must remain within the evolutionary boundaries of the attractor.

Each individual trajectory would thus find its origin in an unstable situation and would actualize various contextual potentialities (typically the first mitosis in embryogenesis, within the same genotype) and would evolve according to the potentialities accessible within the framework of the global structural stability.

Natural selection may reduce variability at the species level, all the while increasing it at the individual level. The more or less tolerable pathologies would rather concern the cells within an individual. Naturally, this conceptual mode perfectly adapts to the above-proposed notion of extended critical situation, as a paradigmatic state of biolons. Besides, it may also be seen as a generalization of Waddington's notion of chreode (Waddington, 1977), by the aspects of genericity and indetermination upon which we have insisted.

Notice that this schema refers solely to biolons, at various levels, given that only biolons are concerned with identities (identities preserved by changes). Orgons would rather constitute, through the energetic exchanges and the functional activities to which they are the locus, the material support to stability and to variation: an orgon is not a trajectory,[8] but may be at the origin of the continuous variation of the individual represented by the trajectory.

Physical paradigms have aided us in formulating these notions, which are not of physical nature proper. We would like to recall, once more, that all the while working within a monist framework, we face various phenomenalities, such at least as they are co-constituted with our living and historical being. Over the course of history, these phenomenalities have been made (partially) intelligible, by their organization using various conceptual structures (and if possible, mathematical ones) – especially over the course of the XXth century, so rich in science, before aiming toward "unification." A synthesis, in fact, is far from obvious; it must be constructed by means of a new (and mathematical) conceptual unification as its long-term objective. As has occurred in physics, the conceptual and technical unification with the biological sciences will be possible only after having specified, we believe, the causal structure and the dynamics specific to each phenomenal level, in its theoretical and experimental autonomy and after having established the conceptual bridges connecting one intelligibility to another, without necessarily, or immediately, reducing one to the other.

---

[8] Let's remind ourselves that a faltering orgon may be replaced by a "prosthesis," an artifact, likely to serve its purpose without decisively affecting the biolon to which it participates (organs within an organism, for instance). This is generally not possible for biolons: in no way can we replace an individual cell.

Chapter 7

Randomness and Determination in the Interplay between the Continuum and the Discrete

Introduction

The idea hinted at in Chapter 5, is that the mathematical structures, constructed for the intelligibility of physical phenomena, according to their continuous (mostly in physics) or discrete nature (generally in computing), may propose different conceptions of nature. We will futher develop this issue here[1].

As a matter of fact, the causal relations, as structures of intelligibility (we "understand nature" by them), are mathematically related to the use of the continuum or the discrete. In particular, as the myth that the world "is" (in the end) discrete or that the universe is, such as the brain or the mechanisms of biological heredity are "discrete algorithms" (the brain as a digital computer, the genetic program), is still largely common, this chapter will challenge these views, in biology and even cognition, from within physics. The focus on randomness will explain our approach to determination by a conceptual duality.

But what discrete (mathematical) structures are we talking about? We believe that there is one clear mathematical definition of "discrete," that we will use in this paper: a structure is *discrete* when the discrete topology on it is "natural." Of course, this is not a formal definition, but in mathematics we all know what "natural" means. For example, one can endow Cantor's real line with a discrete topology (all points are isolated), but this is not "natural" (you do not do much with it, nor do you better understand the reals nor the notion of "continuous functions"). On the other hand, the integer numbers or a digital database are naturally endowed with a discrete

---

[1] A preliminary version of this chapter is in Math. Struct. in Comp. Science, vol 17, n. 2, pp. 289-307, 2007.

topology (even though one may have good reasons to work with them also under a different topological structuring).

In the sequel, the randomness/unpredictability issue in quantum mechanics is going to be discussed. Our approach will then stress that, in the *space-time* of modern microphysics, in no way one may consider the discrete topology as "natural." Our reflection will be based exactly on quantum non-locality and non-separability results that, in our view, propose the exact opposite of an underlying discrete space-time. Indeed, the discrete topology "separates" and "localizes" the elements of mathematical structures, this is its job. Of course, quantum mechanics started exactly by the discovery of a fundamental (and unexpected) discretization of light absorption or emission spectra of atoms (specifically the atom of hydrogen). Then, a few dared to propose a discrete lower bound to the measure of *action*, that is, the product energy×time. It is this physical dimension that has a discrete structure. Clearly, one can then compute, by assuming the relativistic maximum for the speed of light, a Planck length and time. But in no way are space and time thus organized in small "quantum boxes." And this is one of the most striking and crucial features of quantum mechanics: the global and entanglement effects (Bell, 1964; Bohm, 1951; Aspect *et al.*, 1982). These effects suggest the opposite of a discrete, separated organization of space and time and is at the core of its scientific originality.[2] In particular, entanglement motivates quantum computing (as well as our analysis of quantum randomness).

As for the continuum, its role will be stressed in understanding classical determination, as mathematized in the geometry of dynamical systems. As a matter of fact, the notion of randomness lies at the center of dynamic unpredictability: a deterministic system is unpredictable, precisely when it presents random evolutions. In quantum mechanics, this notion is also evoked but in a very different and *intrinsic* way (we will give a more precise meaning to this term). And we will attempt to examine the following issues closely: how does randomness present itself in today's natural sciences? Is there a randomness specific to the various fields of physics? What impact does this eventual differentiation/unity of the notion of physical uncertainties have on the common/scientific concept of randomness? Is it correlated to the various mathematical tools established (continuous vs discrete, for example)?

---

[2] In ongoing work with T. Paul, a quantum physicist (see the papers downloadable from http://www.di.ens.fr/users/longo/), we show how the idea of the topological separability of quanta is the hypothesis underlying the (beautiful) but wrong argument in Einstein *et al.* (1935).

We will not return to the terrain of specifically biological randomness (see Chapter 6) which remains, in our opinion, unexplored. We will firstly reflect upon classical physics, in order to consider the problem of the meaning of randomness in the context of different theoretical frameworks. We will see that dynamical systems and quantum physics independently propose very important, but not identical, notions of *randomness*, giving rise to a different role for probabilities in their various contexts. Our distinguishing criterion will refer to the notion of "epistemicity" as being in contraposition to that of "objectivity," the difference being related to the role of the knowing subject in the construction of scientific objectivity. This is an essential role within modern physics, particularly when it is a question of probability and randomness. In short, we will first state the problem of knowing if a disordered sequence (or more generally, a disordered state) is the effect of chaotic determinism or of pure random processes, of an epistemic or an objective nature. In the absence of very general theorems on the matter (which would contribute to a constitution of objectivity – as is the case for "mixing" random sequences, or Bernoulli's dynamics, that can be demonstrated as being equivalent to – and thus interpreted as – "heads or tails" type sequences; we will return to this), we can develop an argument, which will have the effect of somewhat displacing the problem all the while better highlighting some of the constraints which it implies.

We will develop the thesis that in classical physics, randomness is of an "epistemic" nature, particularly meaning that one will always be able to interpret an apparently random sequence as stemming from a chaotic determinism or as stemming from "pure" randomness, which is analyzable in *statistical terms* (independently on any possible determination). Thus two different approaches are possible, according to the viewpoint one assumes. In contrast, we will call "objective" the randomness specific to quantum mechanics. Our argument will base itself upon two elements justifying this distinction. One is centered upon the role of theoretical determinism, of the "view" (constituted of the proposed theoretical framework) and of measurement. The other, stronger in a certain sense, is founded upon the properties of quantum non-separability and of continuous mathematics, which we will discuss in length.

The fundamental difference between classical and quantum randomness we will explore here may give an idea of one of the difficulties to be found in any attempt to describe a proper "biological randomness," which we mentioned in previous chapters. While waiting for unification (classical/quantum), very expressive theories (classical and relativistic, quantum) have been built and randomness is separately analyzed and successfully applied in the different contexts. In systems biology, in evolution in particular, it may be very hard to propose a suitable theory of randomness, also because this separation of theoretical context seems unsuitable: the least one can say, is that classical interactions (at the phenotypic level typically) are "entangled" with molecular or atomic interactions, and thus, most likely, to quantum effects (in mutations). Thus, randomness in evolution may require an understanding of the *simultaneous action* of both classical and quantum randomness. The discussion in this chapter of the differences, in the understanding of randomness within physics, is not devoid of interest as for possible descriptions of the singularity of biology also concerning this aspect of (in-)determination.

7.1 Deterministic Chaos and Mathematical Randomness: The Case of Classical Physics

Let's therefore begin by analyzing classical randomness. To this end, we assume the idea according to which the *result* of a throw of dice, for example, which we legitimately consider to be random, may be interpreted as such given that the system of equations and of constraints enabling us to describe it (and thus to *determine* it) is (very) sensitive to the initial conditions. Note also that the reverse interpretation is quite different: a dynamic system defined by its equations and presenting a chaotic behavior has an objective character associated with these equations themselves and to their intrinsic properties (it is described/given by physical objects, with their properties, their invariants, their symmetries ... deduced from the equations, see Chapters 3 and 4). In short, from the modern (post-Laplacian) point of view, even this system, a paradigm of randomness, is indeed deterministic, in the sense that a sufficient number (in fact a very great number) of equations could theoretically describe all the forces at play, all of them being *theoretically* well-known (gravitation, various frictions). However, this would tell us very little about its evolution: this classical system is so sensitive to the slightest variation in the boundary conditions, which are, moreover, very numerous, that the mathematical effort of describing the (many) equations determining it will not help us in practice, even not qualitatively. And its evolution remains unpredictable: it is that which leads us to consider it, within a classical framework, as random all the while being deterministic (and chaotic).

A less familiar but simpler example may also be mentioned. Let's consider a double pendulum. We are talking about a pendulum where a second pendulum is articulated on the first, that is a weight is placed at an articulation point of the broken stick of a pendulum. This simple mechanism, perfectly determined by two equations (it has two degrees of freedom), has a chaotic behavior: its trajectories are dense (the weights go everywhere, within the limits of their constraints), it is sensitive to the initial conditions (see the web or Lighthill, 1986, for a forced pendulum). Once more, the system's evolution appears random to all observers, despite the apparent simplicity of the determination. Here we have another case of epistemic randomness, which could in fact also be analyzed in terms of a random sequence (by writing 0 or 1 depending on whether the smaller weight finds itself to the right or to the left after 10 oscillations, for example). On the other hand, a simple pendulum is deterministic and, in principle, predictable (thankfully, Galileo came across a simple pendulum, not a double one, otherwise we would be far from understanding the law of falling bodies, in their basic simplicity).

A final classical example, most relevant given that it triggers all deterministic unpredictability analysis, is the "three-body problem" in their gravitational field, that we have often mentioned. Poincaré demonstrated the impossibility of resolving the system of the nine Newton–Laplace equations, which described its movements in either an elementary and direct manner (by using "simple" functions, let's say) or analytically (by means of convergent series). In doing so, he has enabled us to analyze what we have just done: classical determination may fail to imply predictability. With this result, he has opened the way to the integration-comprehension of classical randomness within the framework of mathematical determination: the "laws" at stake are all clarified by means of equations, however, the evolution remains unpredictable, *thus*, epistemically random. Modern results confirm the scientific relevance of this approach: following Poincaré's geometric approach, Laskar (1990, 1994), has demonstrated that the solar system, our good old planetary system, is chaotic and has given a precise upper bound to predictability. So apart from a few differences regarding the time scales (demonstrable time of unpredictability: 1 million years for Pluto, 100 million years for Earth), conceptually, the situation is not so different for a double pendulum, nor for a throw of dice, from the mathematical viewpoint. In the long term, we could also address it in purely statistical terms, like dice (and, for instance, bet at 2/1 that the Earth will no longer be revolving around the Sun in 500 million years).

The dynamic system, which we are considering, may therefore have a great number of parameters, as do dice, or a medium number of parameters, as does the solar system, or even a very small number of degrees of freedom, as does the double pendulum. We will finally note that this possibility for an alternative (chaotic determinism or randomness to be analyzed in purely statistical terms) is intimately related to the fact that local descriptions of the system are possible: those associated with the underlying equations and their initial conditions, whereas the statistical and probabilistic representations generally involve a global representation of the system. An example of this is the physical behavior of a gas, of which thermodynamics consists precisely in taking global statistical averages – statistical mechanics – of local mechanical behavior.

Let's summarize the two reasons for which we call classical randomness *epistemic*. First, we can choose a purely statistical mathematical approach *or* an analysis in terms of determination (equational). Of course, both approaches, though theoretically equivalent (and there is a sufficient number of theorems demonstrating this equivalence), may be more or less effective or even relevant: normally, we discommend the analysis of planetary evolution in statistical terms just as we do the analysis of dice in equational terms. Each system will have its own best adapted method of analysis. In the case of a double pendulum, the difference is less clear and depends on the aims of the analysis (in any case, this pendulum may very well be used for a little family game of chance ...). Second, the unpredictability of a deterministic system is due to a (classical) physical *principle*: measurement is always an interval. That is, that even if we have a system of equations, which determines a system, point by point (in Euclid's sense of points, or of real numbers as Cantor), only God or an infinite intelligence, Laplace quite soundly tells us, knows the world through (mathematical) points and can thus predict (and retrodict) its future (and passed) states. As far as we, humans, are concerned, our physics uses approached measurements and does so by principle, because, in the worst (or best) of cases, there are classical thermal fluctuations which force approximation (the interval of measurement). Only discrete state machines, such as our digital computers, access *exact* databases, in their discrete, well-separated, topologies, and thus possess predictable evolution, in particular when sequential and, thus, theoretically independent of physical contexts. This was already highlighted by Turing, who views his machine as "Laplacian" (see Longo, 2007 for details and references).

Consider a deterministic system sensitive to the boundary conditions (typically non-linear), thus chaotic.[3] One then has that unpredictability *is the joint result* of this sensitive dependency of the boundary conditions and the *theoretical* properties of classical measurement, an approximated access to the physical world. Classical (and relativistic, of course) theories thus simultaneously give us perfect equational determination (from God's viewpoint, but conceivable even by us, mathematically) and unpredictability. Note that the notion of a chaotic system is purely mathematical: it may be associated with non-linearity of the equations determining a physical process or, more generally, with the conditions mentioned above (sensitivity, topological transitivity, and density by periodic points). While (un-)predictability is a matter of the *relation* between the mathematical system and a physical process. The key role of the mathematical continuum in these frames will be analyzed in Section 7.4. In summary, the following arguments lead us to consider classical randomness as being epistemic:

(1) The possible equivalence of the statistical view and equational determination;
(2) The role of measurement (performed by the knowing subject).

Then, we could propose, for classical dynamics, a "Poincaré thesis":

any classical random process is a trajectory given by a system which is, in principle, deterministic and in a chaotic regime.

This thesis does not necessarily correspond to Poincaré's thinking, but what we can say 120 years after his great theorem. It is, at most, a thesis, because it remains clearly undemonstrable (where is this list of *all* classical processes?), though it is falsifiable.

7.2 The Objectivity of Quantum Randomness

One may argue that in contrast, quantum physics proposes an "objective" randomness, intrinsic to the theory, conceptually and mathematically quite different from classical randomness. This randomness is *intrinsic* inasmuch as it is associated with any operation of measurement, because, in quantum physics, a measurement only returns a *probability* as a result. More specifically, the objective randomness of quantum physics is due, conjunctly (but

---

[3] A more general definition than that suggested uses also the notions of topological transitivity (or density of orbits) and density by periodic points, which we will not develop here, see Devaney (1989).

in each of the cases evoked here quite differently):

(1) To the non-null value of the Planck constant h (this constitutes a lower limit for the product of the possible precisions in the simultaneous measurement of two conjugated variables, namely, for the volume of the phase space, position and impulse);
(2) To the process of measurement specific to quantum physics ("projection of the state vector," which does not depend upon the value of h, all the while being "non-determined" and returning a value as a probability);
(3) To the complex aspect of the wave function: what is added, complex values of the state or wave function, is not what is measured, real numbers; thus, the probability amplitudes, the normalized complex values of the wave function, do not coincide with the probabilities themselves, their squared absolute value (and these probabilities are the probability densities at a position).

Concerning the first two points, the difference with regard to classical measurement is well understood. Quantum theory is centered upon this essential role of *indetermination* in measurement: by the theoretical choice inherent to the approach itself, determination by points (of Euclid–Cantor, as in the mathematics of classical determination) is not possible; such a general determination is theoretically inconceivable, even proscribed, conversely to what is presumed by the mathematics of classical deterministic systems (and which we have called the "Poincaré thesis"). The mathematics of quantum physics get started with Planck's h: it develops through an analysis of quantum states in terms of vectors (state vectors or, in other words, wave functions) within an abstract space of (possibly infinite) dimension (Hilbert space: the space of complex functions the squares of which are integrable). It then comes to propose a linear field where these wave functions are given in terms of complex components. Measurement, which always provides real numbers as a result, is instead associated with an essential loss of information, due to the passing from a complex variable to its absolute value. The physical relevance of the mathematical representation by complex numbers (or phase) is shown by points 2 and 3 above and by phenomena such as quantum interference, where a representation as punctual particles was "classically" expected (*cf.* Young's double-slit experiment). Similar reasons are at the basis of quantum entanglement (an issue to be more closely discussed in the sequel).

So here we already have a few good reasons to consider randomness as intrinsic to the theory: it stems from the measurement, the mathematics, and the evolution of the system (the wave function).

To these it would be necessary to add the intrinsic character of quantum fluctuations (in contrast to classical fluctuations related to temperature, for instance). In measurements, this characteristic is manifested through both the residual energy of the harmonic oscillator (at absolute zero, $0K$) for instance, as well as through the resonance widths in particle theory. In short, $0K$, in classical physics, corresponds to the absolute absence of energy, whereas in quantum mechanics, residual energy is admitted. This plays a remarkable role (speculatively) in cosmology, in the destabilization of the "quantum void" (fundamental energetic state) during the "Big Bang." Regarding the Big Bang, as we already hinted, one may notice, with a few word plays, that this representation resolves at the same time the *enigma* of the Lucretian *clinamen* (of which we have always wondered what could be the origin without any external influence: we find it here in the intrinsic fluctuations) as well as the Leibnizian metaphysical perplexity (*why is there something rather than nothing?*). Well (take it with some humor, but not too much ...), because "nothing," the quantum void, is *energetic* and *unstable* all the while being submitted to these same intrinsic fluctuations: thus, quantum void, fluctuation and Big Bang.

7.2.1 *Separability vs non-separability*

Let's now present our main and possibly new argument for these analyses of randomness (classical vs quantum). Objective quantum randomness appears to be deeply coupled with the properties of *non-separability* (the equation, which describes the evolution of the system, produces an entangled result for the quanta that interacted), as well as with the properties of *non-locality* (the measurement performed upon one of the quanta having interacted produces instantaneous "information" regarding the other's state). It is this indeed which leads us to refute any *local* causal representation, which would seek to account for specific quantum properties. Technically, this refutation is ensured by the Bell inequalities (Bell, 1964), which are in turn validated by the Aspect experiments (Aspect, 1982). In this sense, therefore, the situation of a quantum system may always be considered as *solely* global, without it being possible to reduce it to a combination of local components. This appears to be indissociable from the possibility of establishing causal/deterministic evolution equations. Thus, we will not address here Schrödinger's *local* equation which describes the evolution of a *complex* state vector and which is *beyond* the operation of measurement.

As we already mentioned, the latter concerns another state vector involving the measurement device itself (see, among others, decoherence theory (Zurek, 1991), which corresponds, from the mathematical standpoint, to the passing from complex values to *real* numbers).

This point of view could indicate that in order to account for the situations we have just evoked, rather than resorting to the concepts of "chaotic determination" on the one hand and of "intrinsic randomness" on the other, it would be even more enlightening to use the concepts of "separability" (to characterize the deterministic side) and of "non-separability" (to characterize the intrinsically or objectively random aspect as a participant in the construction of scientific objectivity in quantum physics). In short, it is the possible separation of the different objects which participate in a classical process (each planet of an astronomical system, dice, a simple vs a double pendulum) which enables its mathematical description-determination of observable evolution. In systems sufficiently sensitive to boundary conditions, these objects may evolve in an unpredictable way, although individually always theoretically determined by the dynamics equations. quantum physics, on the other hand, following non-separability (aspect of globality associated with non-locality), definitely confers to randomness a character, which is different, *intrinsic*, we say. Typically, if two flipping coins interact in any classically possible way, then separate while flipping in the air and fall, their analysis may be based on independent probabilities: the observation of one coin sets no limitation on the observation of the other. In contrast to this, two interacting quanta are *entangled*, that is the measurement of one of them sets limitations on the measurement of the other (they cannot be "separated" by observable/measurable properties). In general, thus, a set of n classical random events may be analyzed in statistical terms following classical laws of probability distribution, while quantum observables may violate them, for example when they depend on entangled particles.

In the sequel we will go back to the role of trajectories in a space-time continuum for the classical notion of determination. Now, for a quantum system, there are no hidden variables nor equations, which would determine its evolution – or the "trajectory" of a quantum in an underlying continuum, and this corresponds to the absence of a possible local determination. Note that the so-called "hidden variable" theories do not elude this type of analysis inasmuch as these variables must be considered as non-local, as we mentioned in Chapter 2: they cannot describe only a *local* dependence, on a separated quanta, but also a global dependence (in view of entanglement, see next section).

In conclusion, the classical deterministic chaotic processes (dice, a double pendulum, in contrast to a simple pendulum, but even the solar system) normally also enable another description, in purely statistical terms and this follows classical probabilities. Thus, given that classical physical reality depends on this double description and on the physical (and not mathematical) nature of the limits of measurement, we have insisted on the epistemic character of classical randomness (it would depend on the approach, as is the case for the relationship between statistical mechanics and thermodynamics). On the other hand, in quantum physics, there is no possible double representation: any data, which would enable us to access (to construct) knowledge, any measurement, is a probability, yet of a different nature from the classical one.

Of course, it is question here of an approach, which would need to base itself upon somewhat general theorems in order to be further argued. It, nevertheless, remains that the quantum situation essentially differs from the classical situation regarding the status of the probabilities and randomness it involves and that we cannot avoid taking this situation into consideration.

7.2.2 *Possible objections*

What are the objections we can formulate regarding such a point of view?

(1) *Firstly, that in a way comparable (all the while being different) to the quantum situation, there could exist an intrinsic classical randomness, linked to precisely this collective effect of kinetic energy (temperature) and which would not be reducible to a dynamic system.*
   There are at least two possible responses to this objection. First, it is clear that the role of the initial conditions is determinant: there exists, classically – from the mathematical standpoint, a null set of measurements of the initial conditions, which produces an ordered situation for a gas, for example, all speed vectors are parallel. In contrast to this, quantum non-locality and spontaneous fluctuations necessarily generate a *"clinamen."* Second, for a classical system, the Nernst principle (third law of thermodynamics) states that at as the temperature approaches zero (null kinetic energy, which is conceivable for a classical system) the entropy of the system is null (complete order). Conversely, in a quantum system the indetermination relationships prohibit such a complete order even at this limit (inasmuch as one may conceive attain-

ing it): Planck's h forces a residual energy and, thus, the null kinetic energy is inconceivable. We stress, and more we will do below, that the theoretical possibility of limit cases is crucial in classical approaches: this is the core of Cantor's continuum, a limit construction, and differential equations on it.

These two counter-objections make rather implausible (or even impossible) the transferal of the quantum viewpoint to the classical frameworks and would confirm the meaning of our "Poincaré thesis": a mathematically coherent, yet limited, classical situation which has no quantum meaning. However, a simple example of this, but inverse, limit passage may be given. Classically, it is theoretically conceivable to place a needle on its tip and leave it there for good, if you are very lucky: it is the epistemic nature of your shaking (approximated) hands that make it difficult. Instead, a quantum fluctuation would have it fall in a random direction, under all theoretical circumstances.

(2) *Then, one could more profoundly object that, despite indications to the contrary, the quantum randomness which manifests through these fluctuations is also an effect of the "viewpoint" (it is epistemic) and that the underlying agitation within an environment that one could qualify as sub-quantum (cf. Vigier, Bohm, Halbwachs, Hillion, Lochak) would enable us to account for it, in as much a deterministic way as in the case of chaotic dynamic systems (Louis de Broglie's double solution theory), that is, as a superimposition of a regular and extended solution of Schrödinger equation (wave) and of a singular and much localized solution – particle - corresponding to a non-linear operator – still unknown – inserted within this equation (or Bohm's version of hidden variables theory). This amounts to considering quantum mechanics as only "providing a statistically exact but incomplete description of physical phenomena" (see de Broglie, 1961), an incompleteness which was the perspective of the early Schrödinger or of Einstein – more precisely of EPR (Einstein et al., 1935).*

A possible response to this objection lies in the fact that this environment itself, if one formalizes its effect, intervenes upon quantum magnitudes and their measurements in a global, non-local fashion, as we have noted earlier. Moreover, in order to justify the probabilistic properties of quantum physics and the presence of residual energy, of fluctuations, of virtual particles, etc. these authors referred to a sub-quantum medium or to a new concept of "ether," in principle a continuum, presenting these properties. If one cannot totally discard the possibility of such an approach, it remains that it does not respond to the heuristic principle of conceptual economy, by adding an underlying ether to phenomena. Some may claim that, in the end, this is not determinant. But then, also and foremost, it remains non-local and thus non-classical, within the line of that which has been demonstrated by Bell and Aspect, in view of continuous, yet non-separable, hidden variables.

(3) *One could still argue that this non-locality is similar to the globality presented by a classical system (the gas considered in 1 above, for instance) and therefore reinforces the representation of the effect of a statistical disorder underlying the quantum level itself.*

However, we will respond that the lines of research regarding the unification of physical theories, which lead to the theories of quantum gravitation or to those of superstrings, seem to fit badly with this representation. Particularly, supersymmetries impose, on the one hand, the substitution of a non null dimensionality (strings or p-branes) to the punctual character (therefore of null magnitude) of the structures considered as elementary. On the other hand, the issues regarding the "mass of particles" open up a field of intelligibility and of objectivity which are of a different nature (Higgs fields, or supplementary "compactified" dimensions), and classically inconceivable. All this is a consequence of the rupture from the punctual classical representation and the usual four-dimensional space-time, which raises these "paradoxes."

(4) *Another point to take into consideration and which could nullify our distinction: the passage to the classical limit from quantum mechanics, that is, of reconstructing the classical by making h tend towards 0.*

Now, this passing requires at least double conditions. We only hint at it, as we know that in reality the passing from the quantum to the semi-classical and to the classical is much more delicate than is presented here and that the fundamental problems are still not completely resolved. On the one hand, there must be a canceling out of the Planck constant h (its value going to 0), and, on the other hand, it is also necessary to consider large quantum numbers (the simple canceling of h does not always enable us to construct the classical limit). Once these conditions are assumed, it is tempting to consider them as a way to pass to a classical limit, with its epistemic notion of randomness. In particular, it would be possible, starting from quantum non-separability (in the frame of decoherence theory, for instance, which allows us to understand how the interactions with the environment – the measuring apparatus for instance – destroy entanglement and non-separability), to move away from an objective quantum randomness (if, as we hypothesize, there is a sense to this) and get to a classical epistemic doubling between random representation (purely statistical) and deterministic chaos. This would be compatible with the so-called "quantum chaos," which precisely corresponds to a chaotic classical limit. Thus, many see, in the limit canceling of h, the possibility for a reduction to a dynamical system (thus obtaining the classical equations of mechanics) and, by resorting to large quantum numbers, the possibility for a reduction to epistemic probability. This double passing to the limit is far from being accomplished and constitutes one of the great challenges for the highly sought "unification" between the classical (and relativistic) theories and quantum frameworks.

Having attempted to respond to a few possible strong objections to our approach, there is still another aspect which militates in favor of the objective character of quantum randomness: the profound difference which exists between quantum and classical statistics. This difference, indeed, does not only stem from the non nullity of h, but also stems from the observational properties associated with classical particles in comparison to quanta and to the nature of the symmetries to which bosons and fermions respond (symmetry constraints, which do not exist in the classical framework). The first are discernible whereas the latter are indiscernible, which conduces to different expressions of the statistics to which these entities obey: Fermi–Dirac for fermions (which cannot simultaneously occupy the same quantum state, and which are considered as the matter quanta, endowed with a half-full spin), Bose–Einstein for bosons (which can simultaneously occupy the same quantum state, and which are considered as the interaction quanta, endowed with a full or null spin), while Maxwell–Boltzmann statistics prevail for classical particles. The indiscernibility of the earlier (which is, moreover, related to non-separability) relate back, in our opinion, to the objective character of quantum randomness, whereas the discernability of the latter (the well isolated and "individuated" particles of which we are considering the integral sum of free energy) would refer to the epistemic character of classical randomness.

Moreover, we would like to add another element to this, one of a quite different nature, but one which, in our opinion, has the effect of reinforcing the objective character of quantum probabilities. As we have previously recalled, the domain of physics where probabilities and statistics are the most present is, beyond any doubt, statistical mechanics and thermodynamics, be it at equilibrium or in the study of irreversible phenomena.

To account for these phenomena and because fluctuations (not necessarily quantum) also fill our universe at the same time as there is a certain degree of disorder, Boltzmann was lead to introduce the constant k_B, which bears his name. In a way, the latter measures an entropy, that is, a physical quantity generally related to averages taken from collections. The elements of these collections (a gas, for example, formed of atoms in movement submitted to random shocks) are animated by disordered movements of which the average effect of interaction is represented by the temperature T (in kelvins: K). In the simplest of cases, each of the system's degrees of freedom is associated with an energy which is expressed under the form of: $k_B T/2$. It is to be noted that the other physical theories (gravitational, quantum, electromagnetic) are likely to address isolated elements (two masses, an electron, a photon, in any case, elements of a limited number of degrees of freedom). On the other hand thermodynamics (the domain par excellence of the relevance of the appearance and utilization of k_B) addresses situations where these elements are very numerous (high number of degrees of freedom). This makes it dependent on a statistical mechanics approach (see nevertheless point 2 above for a brief commentary regarding what Louis de Broglie has called the "thermodynamics of the isolated particle").

7.2.3 *Final remarks on quantum randomness*

It is obviously not possible for us to provide a real conclusion: our approach is partly conjectural and to sufficiently support it there lacks general mathematical theorems (on the relationships between classical randomness and chaos). Further developments in the physical theories themselves are also needed which could nurture a vision of contemporary physics thoroughly unified or at least sufficiently objective and discriminating regarding the processes at play in quantum interactions. Currently, the paradigm, which is dominant, which does not necessarily mean definitely established, rather goes in the sense of the conceptions, which we have presented. But the search for a causal interpretation of quantum physics referring for instance to a (continuous) sub-quantum environment is not closed, even if one could consider that, since its inception, its fecundity remains somewhat limited, while progress in other directions is quite rich. We would above all like to avoid finding ourselves in a way overdetermined by a *priori* views too ideological in nature, which would make us want and thus defend a total determinism or, conversely, an essential indeterminism of which in any case the relevance in terms of philosophical implications remains to be demonstrated. Admitting that we, nevertheless, want to go beyond the operational aspect of the scientific approach in order to question the set of significations that it mobilizes (which is also among our preoccupations), it appears to us to be much more interesting to argue for the spatial/energetic/temporal irreducibility due to the non nullity of the h constant. In fact, this irreducibility is the harbinger of considerable developments yet to be established, because it seems in our view to be correlative of these intrinsic probabilities, which we have considered. These are themselves the expression, it appears, of the conditions imposed by indetermination, by the complex character of the wave function, by the non-commutativity, non-separability, etc., in short, by quantum specificities. It seems much more heuristic and fecund to plea and argue for a specificity of living phenomena, which we know to be physico-chemical in the analysis of all of its functionings, but with its mode of existence not reducing itself to the current theories of the inert. A mode of existence which conduces, in order to be understood, to the introduction of concepts as specific as those of metabolism, of normal or pathological, of living or dead, of phylogenesis or ontogenesis, heredity or protension, even of "contingent finality," joint to a proper notion of (entangled?) probabilities.

Finally, it seems important to emphasize the fact that the objective character, which we are assessing here, is a constructed, or *constituted* objectivity. Its constitutive process obviously depends on the standpoint (preparation, measurements, choice of formulations, of convenient mathematical structures and principles, etc.). But once constituted, it becomes independent of the viewpoint in the sense that it resorts to abstract mathematical constants or structures, which, henceforth, prescribe it as much as they describe it. As we have claimed several times during this book, invariance and stability with respect to a change of coordinate system and measure (viewpoint) correspond to constructed objectivity in science.

7.3 Determination and Continuous Mathematics

It is to be observed that, in the preceding sections, we have identified "random" with "unpredictable," in both classical and quantum physics. In the classical case, dynamic unpredictability has provided us with the very definition of randomness, as a consequence of the relationship between mathematical determination on the one hand and physical measurement on the other. In microphysics, we have highlighted that randomness is integrated with the theory itself, that it is in a way the starting point, rooted in its peculiar polarity between knowing subject and object given by means of mathematics, of the preparation of the experiments and of measurement, all conducing to an "unpredictable" result, where only probability is attained. By referring this time to the role of continuous mathematics, we will return to the difference between the two theoretical (and phenomenal) fields, which, relative to randomness, we have separated, respectively, in terms of epistemicity and objectivity (or intrinsic for quantum theory).

When we use the classical viewpoint or tools for the analysis or the production of randomness (we observe a turbulence, throw a coin, a dice), a preliminary analysis of the phenomenon or object is possible: we analyze the irregularities and the stabilities of a fluid, we look at the physical structure, the symmetries of a coin, of a dice. This enables us to ascribe probabilities to the process, which will follow, and to make physically well-founded estimates regarding some elements of this process. In the case of a coin or a dice, the object's *symmetries* and set of physical properties enable us to ascribe probabilities to the occurrence of the various possible events before they take place (1/2, 1/6, respectively). In short, it is possible to *separate* the physical object and its properties from the process, to study it before the measurement relating to the dynamic of interest to us. This is impossible in microphysics: prior to measurement, that is, before the process where randomness will be verified, it is impossible to determine for the physical object its list of properties; particularly, quantum non-separability prevents us from isolating the "properties" of a quantum. In microphysics, the only form of access to the world lies in the measurement of processes. It is impossible to "look" at the photon, the electron, as we can a dice, a coin, independently of its process of production, of evolution and of measurement, where probabilities are intrinsic to observation.

Even when information on possible measurement at the beginning of the process is given, for example, when preparing an electron for the measurement of its spin (by setting a direction, one knows *a priori* that it will be "up" or "down"), there is no underlying classical theory enabling us to conceive the *exact* theoretical determination of this spin, before and independently of measurement, as the result of a determined state or even of a trajectory. It is instead possible to conceive an analysis of the classical dynamics of a coin which, as a solution to movement equations, would describe the coin's exact trajectory, a geodesic along the Euclidean lines of the barycenter and of a point of the edge, for example. The different "hidden variables" approaches in quantum mechanics presumed these classical theories of real underlying trajectories but, as we recalled earlier, they do not elude the non-local aspect of their specifications.

Moreover, the classical theory enables us, conceptually, to go to the limit of measurement. Once again, we very well know that measurement is always an interval, in classical physics; however, the theory of dynamical systems is given within a framework of continuous mathematics, with Euclidean points and trajectories, which are considered as widthless lines. For this reason, as Laplace rightly stated and we already recalled, an infinite and perfect intelligence, knowing the world point by point, could predict everything, including throws of dice. In theory, this boundary continuous framework is, to this day, essential, because the imposing of an *a priori mathematically* finite limit for measurement has no classical physical sense. This then enables us to *conceive* of the complete determination, in particular a trajectory, which begins from a point; it is in this sense that the classical world is deterministic. And randomness, as unpredictability, remains epistemic: it is in the relationship between on the one hand the tool of knowledge and of determination which is mathematics, and on the other hand the object which we presume to be independent and measurable only in a humanly approximated fashion. It is this (presumed) independence which does not occur in quantum physics, where the object and objectivity itself are constituted by the practice of knowledge (preparation of the experiment, measurement and their mathematization, or even their principal mathematical consequence: the quantum object).

In order to better highlight the correlated role of determination and of continuous mathematics within the classical frameworks, in their autonomy relative to the physical object (a preconstituted), let's return to a previous remark. Mathematics performs this passing to the continuous limit in many abstract constructions, which are at the external boundary of the "rational" (ratios between integer magnitudes, for Greek thought). We are thinking, for instance, of the sequences of rational numbers converging towards an irrational, let's say $\sqrt{2}$: this theoretical limit *produces* $\sqrt{2}$, which is not a rational number. It therefore exists within a conceptual universe of actual limits (the geometric construction of $\sqrt{2}$, which so deeply troubled our colleagues in Greece, to the point of leading some of them to the brink of suicide, is the true beginning of mathematics). The understanding stemming from real numbers as Cantor–Dedekind conferred to the mathematics of continua a very robust foundation, as it provided this continuum made of points beyond this world with an immense mathematical effectiveness, stability, and conceptual invariance. It has also enabled the development of modern physics, where the infinite and the passages to the limit play a crucial role well beyond differential calculus. However, it is clear that it does not have any "physical sense" if we are referring to physical measurement, although differential and algebraic calculus, in the continuous framework, are at the center of classical physico-mathematical *determination*. In short, classical determination is a "limit notion" and has sat at the core of the mathematics of continua and mathematical physics for a long time.

The dimensionless points, Euclid told us, are the exact departing points for trajectories, for *widthless* lines, he continued, determined by equations, as Newton, Laplace, and Einstein explained. It is therefore continuous mathematics which enables us to conceive, on the one hand, of theoretically perfect classical determination, and on the other, of the unpredictability of physical evolution, somewhat sensitive to the boundary conditions, following the unavoidable imprecision of physical measurement. Because it is the very idea of a *conceptually* possible continuous substrate, which highlights the approximation of measurement: a universe where the space-time is discrete, digital for instance, would be exact, because it would allow for exact measurements, digit by digit, as *separable* points, and an exact access to information, just as the digital machine accesses its databases.

Let's note that even in turbulence theory, the framework provided by the Navier-Stokes equations is continuous, therefore deterministic, in this limit sense, although specific to highly unpredictable phenomena. And this yields further important difficulties for prediction, even of the theoretical type, due to the absence, even in our day, of proof for the unicity of solutions and therefore for the unicity of the possible trajectory, once given the boundary conditions.

Yet none of this remains for quantum physics, as we argued: it is theoretically impossible to pass to the threshold of possible measurement, to refer to a continuous substrate, be it purely conceptual. The theory begins with Planck's h constant, an inferior boundary of measurement, an indetermination intrinsic to the mathematics of quantum mechanics. There are not, in quantum space, any dimensionless points, possible departures for widthless trajectories: they are proscribed by the theory. In fact, there are no more trajectories whatsoever within space-time, in the classical sense: there lies the radical watershed constituted by quantum physics, after two thousand years of physics of trajectories, from Aristotle to Galileo and Newton to Einstein. In this sense, randomness is intrinsic to the theory, participates in the construction of objectivity, itself becoming "objective."

7.4 Conclusion: Towards Computability

In the previous sections we tried to understand the notion of randomness as unpredictability, in two different theoretical frames (classical dynamics and quantum mechanics). The epistemic nature of classical chaotic dynamics was stressed as based on

- possible alternative understandings (deterministic vs purely statistical),
- an underlying continuum structure that allows us to conceive perfect determination, in the sense of complete infinitary predictability (God, says the theory, who would know the world by Cantorian points, one by one, would be able to predict all future events).

In contrast to this, there are no alternatives to measurement as probability value nor hidden continuous variables in the prevailing interpretation of quantum mechanics, thus the "theoretically objective" nature of quantum randomness (there is no such conceivable God in quantum theories). Moreover, the entanglement effects (non-separability and non-locality) are at the core of the prevailing interpretation of quantum physics and, as stressed above, they contribute to the peculiar nature of quantum randomness. This interpretation is also the basis of modern approaches to quantum computing and cryptography, since deterministic hidden variables are incompatible with current security quantum protocols.

Many computer scientists are familiar with the mathematical definition of randomness proposed by Martin-Löf, related to the Kolmogorof approach and widely developed by Chaitin and others. In short, Martin-Lö f used the classical computability theory to give the notion of "passing any effective statistical test" and defined by this the so called "infinite ML-random sequences." We do not present them here formally (see Calude, 2002), but we just quote one of its major consequences, which highlights the sense of the approach:

an infinite ML-random sequence has no infinite recursively enumerable subsequence.

The meaning of this strong incomputability property should be clear: there is no way to predict or compute infinitely many values of the intended sequence. As a matter of fact, if you had a total recursive function that could output the date and the results of infinitely many lotto or bingo games, you would be very happy (and the sequence would not be random). Infinitely many, of course, otherwise your sequence would be "just" eventually random, which is mathematically the same, for infinite sequences.

How does this mathematical definition relate to our analysis within physics, in terms of dynamics or *processes*? In the mathematical approach, based on computability, there is no reference whatsoever to an underlying physical process generating the sequence, whether it is the lotto, dies or coin tossing. By this, it is truly general and it applies as well to a quantum sequence of, say, an electron's spin-up/spin-down, interpreted as 0s and 1s (of course, this latter sequence is random or unpredictable for rather different reasons, as we said extensively: in particular, the measurement itself contributes to produce the states).

In other words, given a chaotic deterministic dynamics or a quantum phenomenon to which we can associate an infinite sequence of integers, this is ML-random. Clearly, one has to specify how to obtain a sequence from the intended process (e.g. by writing numbers or signs on dies, coins ... by quantum measurement). In the case of a deterministic dynamics in a chaotic regime, once associated a measure and values to the process, the system becomes unpredictable after "enough" time[4]. A fully general and precise statement of this implication is still to be given. That is, how to relate, in full generality, physical randomness to ML or algorithmic randomness. Yet, some recent results in two PhD theses, by Hoyrup and Royas, co-supervised by the second author characterize the dynamics of the processes whose chaotic behavior engenders exactly ML-random sequences. It turns out that in classical dynamics, Birkhoff randomness is equivalent to a (weak form of) ML-randomness in suitably effectivized dynamical systems, see Gacs *et al.* (2009).

The use of recursion theory allowed Martin-Löf to set on robust grounds previous work on randomness, which was done before the invention or a sufficiently widespread use of this theory. The notion of "passing any *effective* statistical test," but also its consequence in terms of inexistence of recursively enumerable subsequences, clarifies two major aspects of randomness

---

[4]Some techniques, such as the analysis of Lyapounov exponents, can give an estimate of this time: for an ago-antagonistic process modeled by the logistic functions with a minimal level of observability of, say, 10^{-15}, one has to wait about 50 iterations, then the kneading sequence becomes random. As a consequence of the work by Laskar on the evolution of the solar system (see references), after 1 million years (not much in astronomical times) the iterated analysis of the position of Pluto whetehr it is still *in* or *out* of the system is a random or unpredictable sequence.

as unpredictability.

First, it provides a frame where one can show that unpredictability is *stronger* than undecidability. As we already mentioned, both classical and quantum unpredictability, which we identified with physical randomness in their contexts, yield ML-randomness as a mathematical notion. In turn, the latter implies a strong form of undecidability, for a sequence: it cannot contain any infinite recursively enumerable subsequence. It is then fair to say that unpredictability is stronger than undecidability, as *non* recursive enumerability, in a context where these two notions can be compared.

Second, Martin-Löf' approach allows us to better understand the interplay between subject/object in (scientific) knowledge. There is neither randomness *in* nature, nor unpredictability. The world is not random nor unpredictable, *per se*: this simply makes no sense. Both randomness and unpredictability pop out in the relation between the world and a knowing subject: in order to predict one needs someone to *pre-dicere* (Latin for *to say in advance*). If nobody says, there is no unpredictability nor physical randomness. In quantum physics, there is even no "physical object" without measurement and mathematics. Now, recursion theory is an eminently linguistic theory: it was born as and it is a matter of algorithms, given in words within formal systems, over sequences of letters or of 0s and 1s. By this, it gives an important contribution to the analysis of physical randomness, when this is defined, as we did, in terms of unpredictability. As a matter of fact, ML-randomness is based on the notion of effectiveness for a *test*, that is, for an activity of someone who wants to test or try to predict the evolution of a sequence. In no way does the relevance of computability in the analysis of this interplay between a knowing subject and the world prove that there is anything intrinsically computational in the world, as many claim. On the contrary, it works on the negative side: it serves only to set a limit to our linguistic (algorithmic) effort *to say something* about the world (to predict) and this by a *"negative"* notion of algorithmic *un*predictability or ML-randomness.

In short, physical (deterministic) unpredictability (at least under the form of Birkhoff randomness, mentioned above) corresponds to algorithimic randomness, a strong form of undecidability or incomputability. Thus, computable processes correspond to predictable physical evolution. This shows that the world is far from being a big digital computer (for a reflection on the "interfaces of incompleteness", see Longo, 2011).

Chapter 8

Conclusion: Unification and Separation of Theories, or the Importance of Negative Results

8.1 Foundational Analysis and Knowledge Construction

This book started off with a foundational analysis of mathematics and physics, which it developed at length. The separation of construction principles and proof principles was the starting point: with this distinction, we highlighted analogies and differences between mathematics and physics, within the framework of a great conceptual unity. If the theoretical analysis of physics was always conducted with a search for solid conceptual and mathematical foundations, it is mainly the analysis of the foundations of mathematics that was motivated by the quest for more or less "absolute" certitudes, from the end of the XIXth century onwards. These motivations, expressed energetically by the founding fathers, were largely due to the crisis stemming from the loss of Euclidean referents in the relationship to space. After 2,500 years, Euclidean unity between the space of the senses and the space of physics crumbled jointly with the end of the absolute coordinates of Newtonian time-space. Likewise, during the XIXth century, an extraordinary boom in mathematics was accompanied by false theorems or false proofs for good theorems, by paradoxes and by contradictions. The quest for "unshakable certainties," to use the words of Hilbert in a letter addressed to Brouwer, was presumably justified. It provided a new mathematical boost in logic, the development of mathematical logic, which was largely centered on arithmetic, and, by these means, provided us with the extraordinary arithmetic machines, which are changing the world.

Today, however, and also thanks to the stubborn efforts of the logicists and formalists of the beginning of the XXth century in providing us with rigor, with effective and unambiguous formalisms, there is no more need, in mathematics particularly, for this certainty to guide foundational analysis.

And our motivation in conducting such reflections is almost opposite to that which still haunts those who seek certainty in the predicative hierarchies of arithmetic induction. In fact, we have tried to go some distance with regard to the constitutive principles of our sciences, within the framework of an analysis we have determined as being of cognitive type. Here, the logical presuppositions were subordinated to the great options of our human (or even animal) practices, but this was mostly done in the attempt to address another: biology. The invariants of action in our space of life are in our view constitutive of these conceptualities, which are expressed by symmetries, order and geodesics, at the center of the conceptual relationships, which we have tried to highlight between mathematics and physics. It is thus that we even consider logic as being constituted by a practice, the dialogical praxis, such as it was initiated in the Greek agora: logical "laws" are the invariants of discourse, of the argumentative dialog. And as with any practice, its role in the constitution of meaning can and must be readdressed by the dynamics of knowledge, particularly when, from a given discipline, one is confronted by another. Our critical analysis of the physico-mathematical foundations is instrumental to the need for new concepts and principles, in biology in particular.

So a foundational analysis, ours, which aims at a "culture of knowledge," which tries to reinitiate discussions concerning not only the constitutive principles of a given body of knowledge, but also that which is implied, the so-called obviousness of common sense. Too often in science, the researcher is immersed in a level of technicality which removes him/her from the sense of his/her origins and from the blinding force of common concepts. Such a researcher is then at risk of remaining imbued with "common sense," deprived of any critical view of his or her own knowledge, and of making rational-universal propositions, which are at risk of being relativized into forms of absolutes.

Our critical methodological attitude does not exclude the commitment to a strong theoretical choice that is motivated, comparative, and supported by proof. But even in such case, critical reflection regarding foundations should be at the center of any scientific activity and accompany the positive scientific proposition, which must not be confounded with the quest for "absolute foundations." By opposing this quest for absolutes, the non logicist mathematician, the one who does not look for foundations in definite and non-human laws, may indeed appreciate the so-called "foundationless" approaches to knowledge. We could even say the same for the physicist who appreciates the relativizing analyses of Einsteinian time and space, as well as the construction of knowledge and of the very objects of knowledge, specific to quantum physics, completely permeated with the relativizing polarity of subject-object. In this discipline, the object, and even its reference system, is co-constituted, between reality and the cognitive subject, in experimental and theoretical activity.

In this perspective, the Wittgensteinian current of thought, among others, quite rightly returned to a problem identified by the second Wittgenstein by developing a very stimulant reflection regarding "knowledge without foundations." It was a true breath of fresh air compared to the obsessive quest for unshakable certainties that is characteristic of platonizing logicism: logical rules outside of the world, or prior to the world, would be normative for mathematics (and for the world). But then and again, and despite the appreciation for such refreshing anti-foundationalism, why insist on the issue of "foundations," why insist upon the distinctions, such as in this book, between "construction principles" and "proof principles," and this, at the center of a foundational proposition, in mathematics or in physics?

In the cases mentioned, the attempt is also epistemological, not only logical; it is also an issue of the quest for an episteme, of conceptual and historical developments, constitutive of the forms of knowledge. It is a question of re-examining knowledge-related and scientific practices, where possible, in order to retrace their dynamics, but mostly, to retrace their constitutive principles, in a "genealogy of concepts," as Riemann observed. In other words, it is an issue of identifying the *sense* of axiomatic, logical propositions, of perspectives which lead to certain theoretical and empirical practices that shape the various sciences. And this must be done not to highlight unshakable certainties, nor to propose absolute or definitive bases. To the contrary, the aim is almost the opposite, as we said earlier and stress once again.

The individuation in mathematics of order or symmetry principles, or the highlighting of the role of the geodesic principle, omnipresent in physics, must be accomplished with the aim of grasping "what underlies," as far as possible, or of finding that which unites entire fields of knowledge; the choices, explicit or not, the constitution of their signification or their "origin," in a sense that is more often conceptual than historical, but which is also historical. We have made this attempt in order to *instigate a dialog* around these very principles and modify them, if it is necessary, and if it can make intelligible other fragments of the world, such as the singular phenomenologies of life.

In our perspective, from Euclid to Riemann and Connes, *common* construction principles provide a foundation for the geometric organization of our relationship to physical space and, actually, for physics. And this, on the basis of the "access" to space and to its measurement. Euclid's rule and compass, Riemann's solid body, and Heisenberg's matrix algebra, employed by Connes, underlie each corresponding theory, all the while considering the radical changes in perspective that each of these approaches proposed: they are the abstract tools for measurement and provide at once the construction principles for their geometries. Likewise, the fact of highlighting that the geodesic principle can make intelligible a development from Newton to Einstein or Schrödinger's equations (derivable by the Hamiltonian optimality, as are Newton's equations) enables us to understand, in a single glance, the power of the theoretical proposition in modern physics, in its successive chapters.

The "foundational" operation which matters to us is therefore the reflection upon the principles of each science; "taking a step aside", looking at them from a distance, to also bring them back as a matter of debate, particularly when addressing other scientific fields. This is what we do, besides, when observing how the phylogenetic "trajectories" of life (and some ontogenetic ones) should no longer be considered as "specific" (geodesic), but rather as "generic" (possibilities of evolution). Conversely, it is rather the living individual who is "specific." In other words, in physics, we recall, the object (experimental) is generic (a falling body, a photon ... may be replaced by any other one: it is an *invariant of an experimental and theoretical practice*) and follows specific (critical) "trajectories," contrarily to what occurs in biology. It therefore constitutes a duality in physics, which enables us to grasp the necessity of a theory of living phenomena which *enriches* the underlying physical principles that also participate in the intelligibility of life. But this duality, generic vs specific, is also a sort of unity, by conceptual symmetry. It is thus the foundational analysis, of physics in this case, which, once performed, enables us to determine the strength and the limits of the physico-mathematical framework, its non-absolute character, the boundaries of its universality, especially when we seek to apply it to biology. So it is a framework that is to be redesigned, beyond its historical domain of construction: the relationship between physics and mathematics.

To conclude, the aim of foundational analysis today is not the same as it was for the founding fathers who were seeking certitudes during an epoch of great foundational crises. This quest, though quite justified at the time, was questioned by many and rightly so. Our objective today is rather to make room for this culture, for this *ethic* of knowledge, which bases itself on the duty of each researcher to make explicit the great *organizing principles* of their own knowledge, to regard them with a critical mind and especially to turn towards other scientific fields in a well thought-out manner. In other fields these principles could be insufficient for understanding and they can be brought up as matter for discussion, even radically, as has been the case with relativity as much as it has been with quantum mechanics relative to classical physics. There is the significance of dynamic universality, specific to scientific knowledge and highly removed from any form of absolute. This knowledge is far also from "relativism": its objectivity is provided by its *constitutive paths*, by its actual friction between the cognizing subject and the world.

8.2 The Importance of Negative Results

As we stressed at length above, the analysis of concepts, conducted on a comparative level if possible, as well as the spelling out, as much as possible, of the philosophical project, should always accompany scientific work. Critical reflection regarding existing theories is at the center of positive scientific constructions, because science is very often constructed *against* the tyranny of existing theories and the supposed autonomy of "facts," which in reality are nothing but "small-scale theories." Science is also often constructed by means of an *audacious interpretation* of "new" (and old) facts; it progresses against the obvious and against common sense (*le "bon sens"*); it struggles against the illusions of immediate knowledge and must be capable of escaping from already established theoretical frameworks with their apparently unshakable principles. For example, the level of mathematical technicity in the geometry of Ptolemaic epicycles constructed from clearly observable facts is very high and was able to produce excellent agreement with astronomical predictions. In order to account for the movements of planets and stars, from the "obvious" immobility of the earth, circles that were added to circles, the centers of new circles, were established with an extraordinary geometrical finesse and gave way to uncountably many "publications," whose authors surely had very high "impact factors" and "quotations indexes." Yet the many writings of the time failed to convince some revolutionary Renaissance thinkers such as Copernicus, Kepler, and Galileo. And, as Bachelard rightly puts it, the construction of knowledge was then founded, as was Greek thought, upon an epistemological severance, which operates a separation with the previous ways of thinking. But it is recent examples that interest us, where the critical view finds expression on a more punctual basis, by means of "negative results." Let's explain.

Access to scientific knowledge is a construction of objectivity, which needs the critical insight of negative results, as the explicit construction of internal limits to current theories and methods. We thus hint at the role of some results which, in logic, in physics or computing, opened up new areas for knowledge, by saying: "No, we cannot compute this, we cannot decide that ..." The idea is that both the sciences of life and of cognition, in particular in connection with mathematics and computing, need similar results, in order to set limits to the passive transfer of physico-mathematical methods into their autonomous construction of knowledge.

When Poincaré was working on the calculi of astronomers, on the dynamics of planets within their gravitational fields, he produced, by purely mathematical means, a great "negative result," as he called it: formal (equational) determination *does not* imply mathematical predictability. The result is "negative," since *one provably cannot predict, or calculate,* the evolution of a planetary system, even if it is formed by only two planets and a sun, despite having a dynamics, which is still perfectly determined by the Newton-Laplace equations. This is the origin of what will later be called "deterministic chaos": systems where determination is compatible with, if not underlying, random evolutions (we have talked of this in chapters 3, 5, and more extensively in chapter 7). It was a true revolution, which destabilized a science that positively expected the great equation of knowledge of the world, as a potentially complete tool for scientific prediction.

Poincaré's result is, of course, important in itself, but its role will be better understood in time, when the *techniques* of the proof (of the "three bodies' theorem") will have spurred a new field of knowledge, the geometry of dynamical systems, of which the applications are quite important within contemporary science. It is not a coincidence if it took 70 years for these techniques to be developed. With the exception of the works by Hadamard and of a few isolated Russian scientists, it took up till the 1950s and 1970s with the Kolmogorov–Arnold–Moser theorem and Ruelle's work: a negative result destabilizes positive expectations and does not necessarily indicate where to go from there. "The new methods" were there in Poincaré's writings, it is true, but the negation of an expectation does not immediately fall within the positivity of science: the delay for applications seems to demonstrate that it is necessary to first assimilate (philosophically) the critical standpoint and the boundaries, which a negative result imposes upon existing knowledge, in order for a new construction of objectivity to follow.

Let's now move more than 40 years later. The critical viewpoint preceded Gödel's 1931 proof of the incompleteness theorem. Gödel did not believe in Hilbert's hypothesis of completeness and decidability of sufficiently expressive formal theories. He thus explored a syntactical variant (through arithmetic) of the liar's paradox, demonstrably equivalent to the coherence of arithmetic: both statements are unprovable, if arithmetic is coherent. The impact of this is also huge. On the one hand, the statement of the theorem, as in the case of Poincaré, surprises and fascinates, on the other, the techniques of proof open up at least one new field: the theory of computability. The notion of Gödelization, the class of recursive functions, defined within the proof, the reflexivity of the meta-theory within the (arithmetic) theory will be at the center of analyses of deduction and effective computation, from the 1930s onwards. The equivalence of the approaches of formal calculi (and deductions), the works of Church, Turing, Kleene, etc., will spur, by means of the methods of proof of Gödel's negative theorem (*one cannot decide ...*), a new discipline, computer science, which is in the process of changing the world: in order to say that one cannot decide, it was necessary to specify what is meant by an "effective procedure of calculus" (and of decision).

In both cases, a theorem which says "no" imposes boundaries upon a form of scientific knowledge (Laplacian determination, formal deduction) and, at the same time, highlights the techniques for progress (quantitative or geometrical methods) or for a better construction of the field thus delimited (effective computability). Because there actually is a difference: Poincaré's new methods already contained, we were saying, the seeds of the geometry of dynamical systems, whereas Gödel's theorem is "only" a (diagonal) theorem of undecidability (Chapter 2), saying nothing about the possible proof of the undecidable statement (actually, on the coherence of arithmetic). We will have to wait for Gentzen (epsilon-induction, 1936), Gödel's 1958 article, or even Girard's normalization theorem in the 1970s in order to have and closely analyze the proofs of coherence. Both theorems therefore set boundaries, but the first also suggests what can be done "beyond," while the second constructs, rigorously, all which is doable "from within" these boundaries.

Let's now recall another immense negative "result" for science. It is not a mathematical theorem, but a change of theoretical viewpoint, following physical experiments. The result consists of the theoretical interpretation of these experiments and the proposition for a radical turnabout in the construction of physical objectivity. In microphysics, *it is impossible to determine,* at the same time, and with as great a precision as one would want, the position and momentum of a particle. Planck, Bohr, Heisenberg impose a change of viewpoint, thus erecting boundaries that are insurmountable for classical physics: the atom *is not* a little planetary system, upon which to apply the classical methods. The classical "field" ends where begins a new analysis based upon the essential indetermination and the correlations of probabilities instead of the classical field and causality leading to the non-locality, the non-separability of quantum phenomena. It is not an issue of the unpredictability of a deterministic system, as for Poincaré, nor of the incompleteness of formal theories (Gödel), but the intrinsic indetermination of a complete system for microphysics.

This breaking in principle shatters the apparent unity of physics, erects a wall between modes of intelligibility within the very field of physics itself: one physical science, centered upon trajectories, from Aristotle to Galileo, to Newton and to Einstein, could tell us very little about a microphysics where quanta do not as such have trajectories across classical nor relativistic space-time. Once this new field of knowledge was constituted, the issue of the unity of science was properly stated (that of physics, at least), this time, in terms of *unification*, rather than in terms of reduction of the relativistic to the quantum field (or vice versa). One hundred years later, the progress is remarkable, but unification is still far from being achieved. In this case, the critical approach is formed at the same time as the analysis of the experiments but, without the total freedom of "hermeneutical" thinking enabling us to first establish limits to the era's perspective, the new construction would be unthinkable; a construction, marked at the onset by a very limited recourse to mathematics in comparison to classical physics. The acritical subscription to the technicity existing in science has its predecessor in the splendid geometry of planetary epicycles, spread across whole volumes that are now completely forgotten.

From the mathematical standpoint, we believe that a great negative theorem (even several theorems) or an epistemic turnaround comparable to that of quantum mechanics is needed, in biology as in cognitive sciences. If we want to see the establishment of a new theoretical field if possible with its own mathematical autonomy (as is the case for dynamics and quantum physics), but even if we want to specify and refine the existing methods (as with Gödel), it is also necessary to target, by means of a critical standpoint, the limits of these methods.

Let's try then to ask: what are the cognitive functions or cerebral (cellular) structures, which are demonstrably ungraspable by formal neural networks and statistical physics? Which boundary is to be set for the analyses of living phenomena in terms of physical criticality (dynamic and thermodynamic)? Is there, in phylogenesis, an indetermination or a randomness, which is specific to living phenomena and comparable, yet different, to indetermination in microphysics (analyses in terms of physical dynamics provide us at best with a deterministic unpredictability)? Which biological phenomenon is non-measurable, in terms of any measure of physical complexity? How can one go beyond the incompleteness of the computational theories of DNA, conceived as a complete (formal-symbolic) "program" for the phenotype (do you remember Hilbert's completeness conjecture?), analyzed in terms of theories which add regulating gene-program over regulating gene-program, not unlike what was done back in the age of epicycles?

8.2.1 Changing frames

Many other results of a "negative nature" may be quoted in science. Let's just mention the various thermodynamic limits (no perpetual movement, no way to reach absolute zero); Kastler (1976) calls them "Actes de renoncement" and refers also to the quantum limits recalled above. Similarly, computer science witnessed a flourishing of negative results. Computational and complexity limits have been shaping the discipline: it is theoretically/practically impossible to compute this or that ... see Harel (2003). Yet, the results we focused on above seem to have provided an epistemological severance as they operated a particularly radical separation with the previous ways of thinking: in computer science, for example, the unfeasibility or limiting complexity results move somewhat along the lines of Gödel's (or Turing's undecidability) theorems, even though the technique and the frame may differ. In short, the results we mentioned above caused a philosophical shock in science and, in particular as for Poincaré's theorem and quantum indetermination, a robust resistance to be "digested" or accepted. In the first case, this was indirectly manifested by the major delay in developing further results along the same lines; in the second, by a persisting minority still now proposing "hidden variables" approaches of a deterministic flavor, in spite of large empirical evidence (at least since Aspect's work on Bell's inequalities in 1980, see Chapter 2).

In the case of the science of the living and cognition, it is possible that the philosophical "resistance" to the required changes in viewpoint, or limiting results, would be even stronger than that which has emerged with regard to unpredictable dynamics, to formal incompleteness and to quantum indetermination: we ourselves constitute living phenomena and, being monists, we want to be within this (physical) world. But the unity of science is a difficult thing to achieve and is not attained by transversally forcing the same methods upon different forms of knowledge, as did the attempt to transfer the little planetary system model to the atom: it doesn't work. First, we would rather need to establish the (causal?) "field" of living phenomena and the boundaries (mathematical boundaries if possible), which define its theoretical autonomy in order to then reach a new synthesis, a unification of "fields", which would probably displace all these boundaries in order to grasp the unicity of the material world (our presumption). Of course, to start off with the available mathematical tools is a good method that is employed by numerous highly valued colleagues. But without the talent for standing back in order to enable critical thinking, as demonstrated by Poincaré and by quantum physicists, it will be difficult to progress much.

The resistance may not only be of a philosophical nature, but may also stem from this "culture of results" more than the "culture of knowledge" we referred to in the previous section, a culture which increasingly claims to completely direct science. The "accountability" obligation, so often required, is of an industrial type and imposes its paradigms: one must beforehand clearly set out the projected methods, the expected results, ... in order to be able, at the end of the project, to compare them with the results effectively obtained.

Scientific objectivity mostly progresses by means of "intelligibility," which may or may not be derived from "positive" results, nor previously accounted methods and objectives. Fundamental research may only be evaluated (and severely so) *a posteriori* and will be fundamental if *it has no foreknowledge of its methods and results*. It is without doubt that applications need a scientific and financial effort: oriented, industrial research is greatly lacking in Europe in particular, but definitely not because of an excess of fundamental research. All the while developing applicative science, it is necessary to maintain a wide platform for perfectly, absolutely independent thought with regard to any conceivable application. What would a corporate director say if the result he/she got from the calculation of the evolution of three bodies within a certain physical field was negative and only to yield repercussions 70 years later? And what if he/she had asked, as an accountable objective, for the exact determination of the position and moment of certain atomic particles? Or if Gödel had been asked to build a digital machine to demonstrate all theorems of combinatorial arithmetic?

The person funding that sort of work would not have been happy with Poincaré, Heisenberg, or Gödel. What would he/she tell the shareholders the following year? Would he/she report a total failure regarding a project of calculus?

Today, and more so than ever, in order to get financing, it is better to propose a computational model for some very precise natural phenomenon, particularly in the fields of biology and cognition, if possible by means of well-established techniques, independently of the target discipline. Proposals to calculate or model, to decide or to determine are certainly at the center of scientific activity, may be very difficult and are highly appreciated (and rightly so). But it would be better, as history teaches us, if, in parallel, we try (and allow researchers) to construct critical views, with their own conceptual frameworks and negative results, that is, with the delimitations that create new fields. And this also requires a *hermeneutic* of scientific knowledge, as was the case for Galilean physics, for relativity and for quantum physics.

As for biology, an ontological monism, we have often repeated, does not imply a monism of theoretical methods, but a scientific unity to be constructed. As within the field of physics, it is possible to aim for unification, once set the relative boundaries, once differentiated the theories, if necessary by means of negative results. Even the mathematics of relativity started off by means of a differentiation of the geometry of the space of senses from that of astrophysics, by a negation: Riemannian geometry *is not* stable by homotheties (enlargements and their opposite). This is the independence from Euclid's fifth axiom: *one cannot transfer* any Euclidean property to distant spaces, the sum of the angles of a triangle, for instance, to triangles of stars (nor to microphysics).

It is therefore necessary to emphasize the role of a critical mode of thinking which does not necessarily aim for a positive result stated beforehand (to calculate or model this or that ...) nor for a result provided by pre-explained methods and expected results, in order for the project to be "accountable" as the Brussels' bureaucrats say, by means of explicit and direct links between promises and results. And it is necessary to maintain an intangible space for a science, which may also produce "non-results" (results that say: "Sorry, but *it is not possible* to calculate, decide, determine ... transfer such or such method, theorem ... "). These results always present a high level of technical difficulty and of originality, but even a controversial idea can be more interesting than a result which is heroic but predictable.

Accountability forces us into "normal science" Kuhn would say, a science which is, *sometimes*, rich in immediate applications. But in the sciences of life and cognition, even more so than in the others, we also need a new theoretical and mathematical view, which would be specific to them. And this, one century and a half after the coming of the theory of evolution, which constituted in its time a revolutionary way of seeing living phenomena, as the only theory truly developed within biology itself and comparable to the great physical theories (relativistic, dynamic, quantum). Thoroughly defining the relative boundaries of the other sciences, physical and mathematical, which claim to be transferable to living phenomena and its cognitive activities, could help to propose it, negatively, and by this help to establish epistemological divisions and new conceptual or even mathematical frameworks.

In this book, we have attempted to provide a few venues, although certainly in an incomplete and preliminary manner: the notion of extended critical situation differentiates the analysis of living phenomena from the current physical theory of criticality, including for the conceptualization of a twofold time dimension specific to biology. Moreover, extended criticality is, in principle, of an infinite physical complexity, as we will recall and further justify below. Indetermination has been described in terms of changes in the very space of evolution, an approach which is foreign to classical physical determination and even to the mathematics of quantum physics. The notion of contingent finality has extended and enriched the usual representations of physical causality, for which the very notion of finality is actually "beyond the subject."

Our idea is that, well beyond our little attempts, and based upon the theoretical originality of Darwinian evolution, only a conceptual or mathematical autonomy of biology could enable the quest for a unity to be constructed by means of physical and physico-chemical theories. And this is the research program we propose in the book: more on a theoretically (and mathematically) autonomous approach to living systems.

8.3 Vitalism and Non-Realism

This epistemological and conceptual separation upon which we insist for a scientific analysis of living phenomena can easily lead to accusations of "vitalism," the latter being slanderous in the hard sciences and recurrent from the moment one proposes an outline of the *theoretical* autonomy of living phenomena. There is no doubt to the monists and materialists we are that there is only matter "out there" and that any form of dualism (mind/body, software/hardware) must be rejected by the scientific analysis of such matter (though the second form of dualism was very convenient for inventing discrete symbolic machines, these human inventions, designed outside of the world, physically brought to exist thanks to technical feats).

Our view is based on the phenomenal evidence, which is brought forth by scientific analysis; it proposes a theoretical separation having one of its great predecessors in the birth of quantum physics, and which surely does not consist of an *ontological* separation. As we do with Darwin's theory of evolution, Varela's autopoiesis, Edelman's notion of degeneracy and its ontogenetic Darwinism (neural, in particular), the Berthoz–Petit theory of action, among others, we first try, if possible, to mathematically identify certain aspects *specific* to the phenomena of life, which seem to be strong invariants that would lend themselves to deeper theorization. The objects of such a view, which would be constitutive of their very objectivity by isolating them, by approaching them using concepts we hope to be appropriate, amount to a few great structural invariants. For example, while looking at time as an operator rather than as a parameter, we put the great biological rhythms at the center of our theorization, as well as we do with retentive and protensive activities (see also Bailly and Longo, 2008), which do not have meaning as such for current physical *theories.*

There must be, of course, behind these great observables, chemical phenomena that molecular biology will reveal one day, just as there are surely acting quanta behind the fall of a falling body in its gravitational field, which the unification of the quantum and relativist fields will presumably reveal one day. However, and for the moment, let's provide ourselves with an autonomous theory of the falling body, of the star which travels in a space of which the curvature correlates gravitation and inertia, and then we will see how to unify the two theories rather than reduce them, as physicists would put it.

In the same spirit and for similar reasons, we propose a qualitative analysis of what we call *extended* criticality, which does not have an equivalent in the theories of physical criticality. The corresponding mathematical methods in physics (renormalization, typically) essentially use the punctuality (as critical value) of the control parameter governing the appearance of the criticality of the observable. Our wager consists in the hypothesis that this critical extension enables us to identify, for the moment on a conceptual level, the difficult play between structural stability and variability in living phenomena, and this from a possibly novel viewpoint: the extended formation and *maintaining* in time and space of a structure that is far from equilibrium, that we could see as individuated and critical, in *any point* of its non-zero measure domain of existence (parameter space). And this, all the while remaining globally and durably stable on a structural level and by involving to these ends the mutual play of regulations and integrations between various levels of organization. If physical theories do not really address this type of biological phenomena, it is for good reasons: the inanimate observed by physicists do not present themselves as such (in relativity, quanta, and classical dynamics). The closest among the "available" physical theories, that of the dynamic of critical phenomena, uses the punctuality (or quasi punctuality) of the field of criticality in an essential (and mathematically powerful) way.

It is in this sense that the *infinite* complexity of the biological, of which we have spoken, relative to the physical, obviously does not mean that we would be unable to describe in physico-chemical terms the stages of biological functioning or that we would have recourse to some biological "ontology" or even worse, to a mystique of the divine infinite. Although within a framework of criticality, there is literally no common measure of biological complexity and of physical complexity *in that the objective complexity of the biological is to be considered as mathematically infinite* relative to objective physical complexity, measured as finite (which does not involve the question of epistemic complexity, see Bailly and Longo, 2003). This is precisely the *mathematical consequence* of our approach in terms of extended criticality, for the following reasons. One of the signatures or elements of universality of physical criticality is given by the divergence of certain values (observable: first derivative, second derivative for example) relative to the control parameter(s): on a *zero measure* set, or on a point or on isolated points, the relevant measurement diverges, without affecting the global mathematical measurement associated with the process. *If our approach makes sense*, the extension of criticality to a non-zero measure set, or an interval, from a mathematical point of view, forces this divergence of which we speak. Besides, as we have thoroughly explained in Chapter 2, infinity is a useful and important concept for the mathematics of physics (typically, for the analysis of phase transitions and critical thresholds): there is nothing magical nor vitalist to it.

The epistemological, or even gnosiological attempted break, centered upon this extension of the range of criticality and its temporal particularities (time as a bi-dimensional manifold), could, in our attempt, constitute a first step towards the construction of one of the possible theoretic frameworks enabling a future unification because, in order to proceed to a *unification*, it is necessary to have at least *two* theories.[1]

Let's return to the analogy with quantum physics. To identify this approach to vitalism (and why not Darwin, Edelman, Berthoz, ...?) corresponds to the accusation, still directed towards the proponents of the standard interpretation of quantum physics, of *not* being *realistic*. Do you not see, says Einstein (see the discussion of EPR in our second chapter), but also says Thom and many other physicists, that there is a *reality*, of which the variables (the causal determinations) are local, of which the properties are separable, even for elementary particles: it may suffice to use a fine enough topology? Well, there is not, others will answer; there are phenomena of non-locality, of non-separability and, moreover, the properties of probability in measurement are intrinsic (objective, we said in chapter 7). The epistemological break created by this theoretical action is enormous, a real theoretical "dualism." But the acknowledgment of the autonomy of quantum physics has provided us with a fascinating theory having surprising practical spin-offs. It also gives us today the possibility of considering a unification with the relativistic theory of gravitation, though at the cost of a change in the theory of the relevant space (non-commutative geometry), or even of quantum objects (string theory). This will not be inconsequential, of course, relative to the geometry of the relativistic field itself.

Currently, some even believe that the methodology of quantum physics, the type of view that it takes, even the geometry of space which it proposes, could be exported towards other theoretical frameworks: such as the work by Mugur-Schachter (on the method of conceptualization) and that of Girard, whose new geometry of proof is permeated by non-commutative geometry. Bitbol rightly points out that the subject/object intrication in the cognitive sciences would have a lot to learn, on the level of conceptual practices, from intrication such as it appears technically (theoretically and experimentally) in quantum physics. Without the hermeneutical audacity of a few physicists before the Second World War (afterwards, it became extremely difficult to propose radically new points of view; not that it is impossible, just more difficult), we would still be trying to transpose the mathematical methods for Poincaré's three-body dynamic (another revolution within his field of classical dynamics, digested with great delay: can't we positively predict what is determined?) to the three bodies of helium.

---

[1] More recent work on "anti-entropy" goes further in this direction (Bally and Longo, 2009).

The impossibility of such a type of transferal leads us to ponder upon the universality of the mathematical tool. Mathematics, *once it is constituted*, is doubtlessly "generic" in the sense that, by means of its maximal invariance and conceptual stability among the forms of knowledge construction, it is no longer "specific" to a given domain of nature. To put it in phenomenological terms, after transcendental constitution, which is contingent, it becomes transcendental, that is, it becomes idealities. On the other hand, there are distinct ways in which to use mathematics according to the field of study, and the history of physics demonstrates that when needed, a discipline can invent mathematics that does not yet exist, be it integrated later on within a corpus having dynamic boundaries while being remarkably unitary (recent examples: Dirac's delta and the theory of distributions, path integrals, etc.). One could think that the field of biology would lead us to invent yet "inexistent" mathematics which would later on be justified on its own merits by mathematicians (such could be the case for a non-null dense set of critical points, for example).

It is in this sense that we say that mathematics, relative to issues of knowledge, is not transferable as such, except in some rare occasions. One example, among these exceptions, is the transferal of Girolamo Cardano's complex algebraic numbers to quantum conjugation: well-established mathematical structures and concepts (mathematical *idealities*, in short) contribute to the *construction of a new physical objectivity* (in fact, of quantum *objects*), without co-constituting itself with it. For once, but there are other exceptions, not only definitions, but a robust theory and some of its deep results may be as such applied to a new scientific area.

But normally, mathematics is shaped or reshaped around a new phenomenality, all the while co-constituting its theoretical objects, as was the case for infinitesimal calculus, Riemannian geometry, the geometry of dynamic systems, etc. Of course, starting off with the available tools is also a good strategy. But then, as we said, it is necessary to have the talent for detachment of a great thinker such as Poincaré who, while trying to solve the Lindstedt problem with the tools available to him, noticed that the difficulty was not technical, but intrinsic: it required a change of viewpoint. And this will enable him to demonstrate a great negative result, which changed the mathematics of dynamic systems. As far as we are concerned, aware of the discrete charm of mathematical technicity which, by the way, we have experienced, we preferred to draw first some conceptualizations of living phenomena that are permeated by the physico-mathematical (by dualities, symmetries ...), but that are *theoretically* well separated from current physical theories (and some mathematics is starting to follow, see some recent papers of ours). With, in the background, the solid guarantee of a materialism and a monism (ontological) which enabled us to look, far beyond artificial separations, as those of the true (ontological) dualisms, between life and inert, syntax and semantics, software and hardware, which have been for too long presumed, as such, to be a model of human thought and brain, mind and body.

For over 100 years, theoretical separation has been at the center of scientific construction, beyond metaphysical dualism. Unifications follow and they may also be of a great conceptual and technical difficulty because they in fact require the invention of new theoretical frameworks.

8.4 End and Opening ...

So here we are at a (provisional) stage of the process which has led us from the re-examination of the foundations of mathematics and of physics to the search for structural and conceptual regularities in the field of the biological sciences. Throughout this process, we have examined the new views which these analyses could provide in terms of the various causality regimes involved by theorization and explanation as well as in terms of the relevant invariants likely to be highlighted and participating in the construction of objectivity. Of course, this process does not claim to be exhaustive, but it provides, in our opinion, a renewed view of the elements of a contemporary philosophy of nature and of the conditions of functioning and development of human cognition throughout its theorizations and conceptualizations relative to mathematics and to the natural sciences.

The sometimes diagonal form in which these ideas have been presented does not only reflect the scientific histories and sensibilities which differ according to the original disciplines of the authors. On the one hand, they correspond to real dialogs having taken place between them and which have contributed to shaping the respective approaches of each author towards a common viewpoint. On the other hand, our text should contribute to bringing a less unilateral view of the problems addressed such as to enable, we hope, the reader to better participate in the debate according to his or her own reflections and areas of interest.

Regarding the matter as such, besides the reactualized examination of the great cognitive and interpretative controversies that marked the history and conceptualization of mathematics and physics, we attempt, albeit in a very limited fashion, to extend this type of discussion to the biological sciences for which the theorizations still remain partial and fragmented. We do this not in order to impudently propose a complete biological theory, nor to transpose methods constructed elsewhere upon it, but rather in order to attempt to identify certain issues, which seem relevant and certain objectivities, which could prove to be more robust. In fact, we are aware that in this field the speculative aspect tends to largely supersede the established technical aspect but, in our view, this could provide the opportunity to freshly engage in a theoretical and conceptual debate. Such an ongoing debate could finally prove to be fructuous for a deep understanding of the "living state of matter" as well as for the analysis of human cognition, in a rich interaction with mathematics and physics and their common *principles of construction* ("da capo").

Bibliography

Aceto L., Longo G. and Victor B., (eds.) *"The difference between Sequential and Concurrent Computations,"* special issue of: **Mathematical Structures in Computer Science**, Cambridge University Press, no. 4–5, 2003.

Adler R. L., **Topological entropy and equivalence of dynamical systems**, Memoirs of the American Mathematical Society no. 219, American Mathematical Society, Providence, RI, 1979.

Alligood K., Sauer T. and Yorke J., **Chaos: an introduction to Dynamical Systems**, Springer, New York, 2000.

Amari S. and Nagaoka H., **Methods of Information Geometry**, American Mathematical Society Translations, vol. 191, American Mathematical Society and Oxford University Press, 2000.

Ameisen J.C., **La sculpture du vivant. Le suicide cellulaire ou la mort créatrice**, Éditions du Seuil, Paris, 1999.

Anandan J., *"Causality, Symmetries and Quantum Mechanics"* **Foundations of Physics Letters**, vol. 15, no. 5, pp. 415–438, 2002.

Aspect A., Grangier P. and Roger G., *"Experimental Realization of the Einstein–Podolsky–Rosen–Bohm Gedankenexperiment: A New Violation of Bell's Inequalities,"* **Physical Review Letters** vol. 49, p. 91, 1982.

Asperti A. and Longo G., **Categories, Types and Structures**, MIT Press, Cambridge, MA 1991.

Aubin J.P., **Viability theory**, Birkhauser, Boston 1991.

Auger P., Bardou A. and Coulombe A., *"Simulation de different mécanismes électrophysiologiques de la fibrillation ventriculaire"*, in: **Biologie théorique** (Y. Bouligand, ed.), pp. 197–209, CNRS Editions, Paris, 1989.

Babloyantz A. and Destexhe A., "*Nonlinear analysis and modelling of cortical activity*", in: **Mathematics applied to biology and medicine** (J. Demongeot and V. Capasso, eds.), Wuerz Publishing, Winnipeg, 1993.

Badiou A., *Le Nombre et les nombres*, Éditions du Seuil, Paris, 1990.

Bailly F., "*Niveaux d'organisation, changements de niveaux, finalité*", **Philosophica**, vol. 47, p. 31, 1991a.

Bailly F., "*L'anneau des disciplines*", **Revue Internationale de Systémique**, vol. 5, no. 3, 1991b.

Bailly F., "*Sur les concepts d'autonomie et d'hétéronomies dans les disciplines scientifiques et leur extension métaphorique*", **Revue Internationale de Systémique.**, vol. 13, no. 3, p. 253, 1998.

Bailly F., "*Construction d'objectivité et statut de l'objet-scientifique en physique*", **La Nuova Critica**, vol. 36, 2000.

Bailly F., "*About the emergence of invariances in physics: from "substantial" conservation to formal invariance*", in: **Quantum Mechanics, Mathematics, Cognition and Action** (M. Mugur-SchÃd'chter and A. van der Merwe, eds.) Kluwer Academic Publishers, Dordecht, 2002.

Bailly F., "*Le "concept–scientifique" reste-t-il toujours un "concept"?*" **La Nuova Critica**, no. 43–44, 2004.

Bailly F., Gaill F. and Mosseri R., "*La fractalité en biologie: ses relations avec les notions de fonction et d'organisation*", in: **Biologie théorique** (Y. Bouligand, ed.), pp. 75–93, CNRS Editions, Paris, 1989.

Bailly F., Gaill F. and Mosseri R., "*Orgons and biolons in theoretical biology: Phenomenological analysis and quantum analogies,*" **Acta Biotheoretica**, vol. 41, p. 3, 1993.

Bailly F., Gaill F. and Mosseri R., "*Morphogenèse et croissance biologique: un modèle dynamique simple pour le poumon*, in: **La biologie théorique à Solignac** (H. Vérine, ed.), pp. 65–94, Polytechnica, Paris, 1994.

Bailly F. and Longo G., "*Objective and Epistemic Complexity in Biology*" in: **Proceedings of the International Conference on Theoretical Neurobiology**, (N. D. Singh, ed.), pp. 62–79, National Brain Research Centre, New Delhi, 2003.

Bailly F. and Longo G., "*Space, Time and Cognition: From the Standpoint of Mathematics and Natural Sciences*" in: **Mind and Causality**, (Peruzzi, ed.), pp. 149–199, John Benjamins Publishing, Amsterdam, 2004 (Published in French as: "*Espace, temps et cognition. A partir des mathématiques et des sciences de la nature,*" **Revue de Synthèse**,

Presses de la rue d'Ulm, Paris, vol. 124, 2003).

Bailly F. and Longo G., "*Incomplétude et incertitude en mathématiques et en physique,*" in: **Il pensiero filosofico di Giulio Preti**, (Parrini and Scarantino, eds.), pp. 305–340, Guerini e Associati, Milan, 2004.

Bailly F. and Longo G., "*Extended Critical Situation,*" **Journal of Biological Systems**, vol. 16, no. 2, pp. 309–336, 2008.

Bailly F. and Longo G., "*Biological Organization and Anti-Entropy,*" **Journal of Biological Systems**, vol. 17, no. 1, pp. 63–96, 2009.

Bailly F., Longo G. and Montevil M., "*Geometric schemes for biological time*". To be published, 2010.

Bailly F. and Mosseri R., "*Symétrie*", in: **Dictionnaire d'histoire et de philosophie des sciences**, Presses Universitaires de France, Paris, 1999.

Bak P., Tang C. and Wiesenfeld K., "*Self-organized criticality,*" **Physical Review A**, vol. 38, pp. 364–374, 1988.

Barreau H. and Harthong J. (eds.), **La mathématique non standard**, CNRS Editions, Paris, 1989.

Barwise J. and Moss L., **Vicious Circles: on the mathematics of non-wellfounded phenomena**, CSLI Publications, Stanford University, 1996.

Bell J., **A Primer of Infinitesimal Analysis**, Cambridge University Press, 1998.

Bell J.S., "*On the Einstein–Podolsky–Rosen Paradox*", **Physics**, vol. 1, p. 195, 1964.

Bennequin D., "*Questions de physique galoisienne*", in: **Passion des forms, à René Thom**, (M. Porte, ed.), pp. 311–410, ENS Editions, Fontenay-Saint Cloud, 1994.

Bernard-Weil E., **Stratégies paradoxales en bio-médecine et sciences humaines**, L'Harmattan, Paris, 2002a.

Bernard-Weil E., "*Ago-antagonistic Systems,*" in: **Quantum Mechanics, Mathematics, Cognition and Action**, (M. Mugur-SchÃďchter and A. Van der Merwe, eds.) Kluwer Academic Publishers, Dordecht, 2002b.

Berthoz A., **The Brain's Sense of Movement**, translated by Giselle Weiss, Harvard University Press, Cambridge, MA, 2002 (first published as: **Le Sens du Movement**, Editions Odile Jacob, Paris, 1997).

Berthoz A. and Petit J.-L., "*Repenser le corps, l'action et la cognition à la lumière des neurosciences*", **Intellectica**, no. 36–37, 2003.

Bitbol M., **L'aveuglante proximité du réel**, Champs-Flammarion, Paris,

2000a.

Bitbol M., **Physique et philosophie de l'esprit**, Champs-Flammarion, Paris, 2000b.

Bohm D., *"The Paradox of Einstein, Rosen and Podolsky,"* **Quantum Theory**, pp. 611–623, Dover Publications, New York, 1951.

Bohm D., **La plénitude de l'univers**, Paris, Le Rocher, 1987.

Boi L., **Le problème mathématique de l'espace**, Springer, Heidelberg, 1995.

Boi L., *"Theories of Spacetime in Modern Physics,"* Synthese vol. 139, no. 3, pp. 429–489, 2004a.

Boi L., *"Geometrical and topological foundations of theoretical physics: from gauge theories to string program,"* **International Journal of Mathematics and Mathematical Sciences**, vol. 34, pp. 1777–1836, 2004b.

Bouligand Y., **La morphogenèse: de la biologie aux mathématiques**, Maloine, Paris, 1980.

Bouligand Y., (ed.), **Biologie théorique, Solignac 1987**, CNRS Editions, Paris 1989.

Bourgine P. and Stewart J., *"Autopoïesis and Cognition,"* **Artificial Life**, vol. 10, no. 3, pp. 327–345 2004.

de Broglie L., **Introduction à la nouvelle théorie des particules de M. Jean-Pierre Vigier et de ses collaborateurs**, Gauthier-Villars, Paris, 1961.

Brouwer L., *"Consciousness, Philosophy and Mathematics,"* in: **Collected Works, Philosophy and Foundations of Mathematics Vol. 1** (A. Heyting, ed.), North-Holland, Amsterdam 1975.

Buiatti M., **Lo stato vivente della material**, UTET Libreria, Torino, 2000.

Burri P.H., J. Dbaly and Weibel E.R., *"The postnatal growth of the rat lung, 3,"* **Anatomical Rocord**, vol. 178, p. 711, 1974.

Butterworth B., **The mathematical brain**, MacMillan, London 1999.

Cadiot P. and Visetti Y.-M., **Pour une théorie des formes sémantiques; motifs, profils, thèmes**, Presses Universitaires de France, Paris, 2001.

Calude C., **Information and randomness**, Springer-Verlag, Berlin, 2002.

Canguilhem G., **La connaissance de la vie**, Broché, Paris, 2000.

Cavaillès J., **Méthode axiomatique et formalisme**, Hermann, Paris, 1981.

Chaline J., **Les horloges du vivant**, Hachette, Paris, 1999.

Châtelet G., **Les enjeux du mobile**, Éditions du Seuil, Paris, 1993 (Published in English as: **Figuring Space: Philosophy, Mathematics, and Physics**, Kluwer Academic Publishers, Dordrecht, 1999).

Church A., *"A set of postulates for the Foundation of Logic"* **Annals of Mathematics**, vol. 2, no. 33, pp. 348–349 and no. 34 pp. 839–864, 1932-3.

Church A., *"A formalisation of the simple theory of types"* **Journal of Symbolic Logic** vol. 5, pp. 56–58, 1940.

Cohen P.J., **Set Theory and the Continuum Hypothesis**, W. A. Benjamin, New York, 1966.

Cohen-Tannoudji G. and Spiro M., **La matière-espace-temps**, Fayard, Paris, 1986.

Connes A., **Géométrie noncommutative**, Inter éditions, Paris, 1990. (Published in English as: **Noncommutative Geometry**, Academic Press, San Diego, CA, 1995).

Curry H. B., Feys E., Curry H. B., Hindley J. R. and Seldin J., **Combinatory Logic vol. I**, North-Holland, Amsterdam, 1968

Curry H. B., Feys E., Curry H. B., Hindley J. R. and Seldin J., **Combinatory Logic vol. II**, North-Holland, Amsterdam, 1972.

van Dalen D., *"Brouwer's dogma of languageless mathematics and its role in his writings"* **Significs, Mathematics and Semiotics** (E. Heijerman ed.), pp. 33–44, Nodus Publikationen, Amsterdam, 1991

Dehaene S., **La bosse des Maths**, Editions Odile Jacob, Paris, 1997 (Published in English as: **The Number Sense**, Harvard University Press, Cambridge, MA, 1997).

Dehaene S., Izard V., Pica P. and Spelke E. *"Core Knowledge of Geometry in an Amazonian Indigene Group"*, **Science**, vol. 311, pp. 381–384, 2006.

Delamotte B., *"A hint of renormalization,"* **American Journal of Physics** vol. 72, p. 170, 2004.

Delattre P. and Thellier M., **Elaboration et justification des modèles**, Maloine, Paris, 1980.

Demongeot J., Estève F. and Pachot P., *"Chaos et bruit dans les systèmes dynamiques biologiques,"* in: **Biologie théorique** (Y. Bouligand, ed.), CNRS Editions, Paris, 1989.

Devaney R. L., **An introduction to chaotic dynamical systems**, Addison-Wesley, Reading, MA, 1989.

Dorato M., **Time and Reality**, CLUEB, Bologna, 1995.

Doridot F. and Panza M., *"A propos de l'apport des sciences cognitives à la philosophie des mathématiques,"* **Intellectica**, no. 39/2, pp. 263–287

2004.

Duff M.J., "*Kaluza-Klein Theory in Perspective*," in: **Proceedings of the Symposium 'The Oskar Klein Centenary'**, (Lindström, U., ed.), pp. 22–35, World Scientific, Singapore, 1994.

Edalat A., "*Domains for Computation in Mathematics, Physics and Exact Real Arithmetic*," **Bulletin for Symbolic Logic**, vol. 3, no. 4, pp. 401–452, 1997.

Edelman G., **Neural Darwinism**, Basic Books, New York, 1987.

Edelman G. and Tononi G., **A Universe of Consciousness. How Matter Becomes Imagination**, Basic Books, New York, 2000.

Einstein A., Podolsky B. and Rosen N., "*Can Quantum Mechanical Description of Physical Reality Be Considered Complete?*" **Physical Review** 41, p. 777, 1935.

Feferman, S., "*Weyl Vindicated: "Das Kontinuum" 70 Years Later,*" **Proceedings of the Cesena Conference in Logic and Philosophy of Science**, vol. 1, Cesena, 7–10, Bologna, 1987.

Feferman S., "*Penrose's Goedelian argument*," manuscript, Department of Mathematics, Stanford University, 1995.

Follesdal D., "*Gödel and Husserl*," in: **Naturalizing Phenomenology** (J. Petitot, F. Varela, B. Pachoud and J-M. Roy, eds.) Stanford University Press 1999.

Foster J.G., *Physics with Two Time Dimensions*, Senior Thesis, Duke University, Durham, 2003.

van Fraassen B., **Laws and Symmetry**, Oxford University Press, 1989.

Frege G., **The Foundations of Arithmetic: A Logico-Mathematical Enquiry into the Concept of Number**, translated by J. L. Austin, Northwestern University Press, Evanston, 1980.

Fuchs C. and Victorri B., **La Polysémie, construction dynamique du sens**, Hermès, Paris, 1996.

Gacs P., Hoyrup M. and Rojas C., "*Randomness on Computable Metric Spaces: A dynamical point of view*," in: **26th International Symposium on Theoretical Aspects of Computer Science** (STACS) 2009.

Girard J.-Y., "*Linear Logic*" **Theoretical Computer Science**, vol. 50, pp. 1–102, 1987.

Girard J.-Y., "*Locus Solum*," special issue of: **Mathematical Structures in Computer Science**, Cambridge U.P., vol. 11, no. 3, 2001.

Girard J.-Y., Lafont Y. and Taylor P., **Proofs and Types**, Cambridge University Press, 1990.

Glass L. and Mackey M., **From clocks to chaos. The rhythms of life**, Princeton University Press, 1988.

Gödel K. *"Russell's mathematical logic"* in **The philosophy of Bertrand Russell** (P.A. Schlipp, ed.), Northwestern University Press, Evanston, IL, 1944; reprinted in **Philosophy of mathematics; selected readings** (Benacerraf and Putnam, eds.), Prentice Hall, 1964.

Gödel K. *"What is Cantor's Continuum Problem?,"* **Amer. Math. Monthly**, 54, 1947; reprinted in **Philosophy of mathematics; selected readings** (P. Benacerraf and H. Putnam, eds), Prentice Hall, Englewood Cliffs, NJ, 1964.

Gould S.J., *Allometry and size in ontogeny and phylogeny*, **Biological Review**, vol. 41, p. 587, 1966.

Gould S.J., **Ever since Darwin**, Norton, New York, 1977

Gould S.J., **Wonderful Life: The Burgess Shale and the Nature of History**, Harvard University Press, Cambridge, MA, 1989.

Gould S.J., **Full House: The Spread of Excellence From Plato to Darwin**, Harmony Books, New York, 1998.

Green M., Schwarz J. and Witten E., **Superstring theory**, Cambridge University Press, 1988.

Haken H., **Synergetics**, Springer, Heidelberg, 1978.

Harel D., **Computers Ltd.: What They Really Can't Do**, Oxford University Press, 2003.

Harrington L. *et al.* (eds.), **Harvey Friedman's Research on the Foundations of Mathematics**, North-Holland, Amsterdam, 1985.

Harthong J., *"Eléments pour une théorie du continu,"* **Astérisque** vol. 109–110, p. 235, 1983.

Heath T.L., **The Thirteen Books of Euclid's Elements**, Cambridge University Press, 1908.

Herrenschmidt C., **Les trios écritures**, Gallimard, Paris, 2007.

Hertz J., Krogh A. and Palmer R., **Introduction to the Theory of Neural Computation**, Addison-Wesley, 1991.

Hilbert D., **Grundlagen der Geometrie**, Teubner, Leipzig, 1899. (Published in English as **Foundations of Geometry**, translated by L. Unger, Open Court, La Salle, 1971).

Hill E.L., *"Hamilton's Principle and the Conservation Theorems of Mathematical Physics,"* **Reviews of Modern Physics**, vol. 23, no. 3, p. 253, 1951.

Horsfield R., *"Postnatal growth of the dog's bronchial tree,"* **Respiratory Physiology**, vol. 29, p. 185, 1977.

Horsfield R., "*Morphology of branching trees related to entropy*", **Respiratory Physiology**, vol. 29, p. 179, 1977.

Husserl E., **The Origin of Geometry**, 1933. (transl. in English, University of Nebraska Press, 1989).

Husserl, E., **Storia critica delle idee**, (translation of **Part I of Erste. Philosophie** by Piana, G., Guerini e Associati: Milan, 1989.

Jean R., **Phyllotaxis: a systemic study in plant morphogenesis**, Cambridge University Press, 1994.

Jeong H., Tombor, B., Albert,R., Ottvai, Z.N. and Barabasi A-L., "*The large scale organization of metabolic networks*," **Nature**, vol. 407, p. 651, 2000.

Johnstone P., **Topos Theory**, Academic Press, New York, 1977.

Kaku M., **Hyperspace**, Oxford University Press, 1994.

Kant E., **Critique de la raison pure**, Paris, Presses Universitaires de France, 1986a.

Kant E., **Opus postumum**, Paris, Presses Universitaires de France, 1986b.

Kastler A., **Cette étrange matière**, Stock, Paris, 1976.

Kauffman S., **The origins of order**, Oxford University Press, 1993.

Kauffman S., **At home in the Universe**, Oxford University Press, 1995.

Lakoff, G. and Nunez, R., **Where Mathematics Comes From: How the Embodied Mathematics Creates Mathematics**, Basic Books, New York, 2000.

Lambek J. and Scott P.J., **Introduction to higher order categorical logic**, Cambridge University Press 1986.

Largeault J., **Logique mathématique. Textes**, Armand Colin, Collection U, Paris 1972.

Laskar J., "*The chaotic behaviour of the solar system*", **Icarus**, vol. 88, pp. 266–291, 1990.

Laskar J., "*Large scale chaos in the Solar System*", **Astronomy & Astrophysics**, vol. 287, L9–L12, 1994.

Lassègue J., **Alan Turing**, Les Belles Lettres, Paris, 1998.

Lautman A., **Essai sur l'unité des mathématiques**, UGE, Paris, 1977.

Lebowitz J., "*Microscopic Origins of Irreversible Macroscopic Behavior*," **Physica A** vol. 263, pp. 516–527, 1999.

Lecointre G. and Le Guyader H., **Classification phylogénétique du vivant**, Belin, Paris, 2001.

Lefèvre J., "*Teleonomical optimization of a fractal model of the pulmonary arterial bed*", **Journal of Theoretical Biology**, vol. 102, p. 225, 1983.

Lesne A., **Approches multi-échelles en physique et en biologie**, **Habilitation à diriger la recherche**, Universitaire Paris VI, 2003.

Lichnerowicz A., **Théories relativistes de la gravitation et de l'électromagnétisme**, Masson, Paris, 1955.

Lighthill J., "*The recent recognized failure of predictability in Newtonian dynamics*," **Proceedings of the Royal Society of London A** 407, pp. 35–50, 1986.

Lobachevskij N., **Nouveaux principes de la Géométrie**, 1856.

Longo G., "*Set-Theoretical Models of Lambda-calculus: Theories, Expansions, Isomorphisms,*" **Annals Pure Applied Logic** vol. 24, pp. 153–188, 1983.

Longo G., "*Some aspects of impredicativity: Weyl's philosophy of mathematics and today's Type Theory*" in: **Logic Colloquium 1987: Studies in Logic and the Foundations of Mathematics** (H.-D. Ebbinghaus *et al.* eds.), vol. 129, pp. 241–274, North-Holland, Amsterdam, 1989.

Longo G., "*De la cognition à la géométrie*", **Intellectica** no. 25, 1997a.

Longo G., "*Géométrie, Mouvement, Espace: Cognition et Mathématiques. À partir du livre "Le sens du mouvement""*, par A. Berthoz, Odile-Jacob, **Intellectica**, no. 25, 1997.

Longo G., "*The Mathematical Continuum, from Intuition to Logic*" in: **Naturalizing Phenomenology** (J. Petitot, F. Varela, B. Pachoud, J-M. Roy, eds.), Stanford University Press 1999.

Longo G., "*Mathematical Intelligence, Infinity and Machines: beyond the Gödelitis,*" **Journal of Consciousness Studies**, vol. 6, no. 11–12, 1999b.

Longo G., "Mémoire et objectivité en mathématiques" in: **Le réel en mathématiques**, (P. Cartier and N. Charrauds, eds.), Presses de la rue d'Ulm, 2003, actes du Colloque de Cerisy, 1999c.

Longo G., "Cercles vicieux, Mathématiques et formalisations logiques," Conférence Invitée, parue dans **Mathématiques, Informatique et Sciences Humaines**, n. 152, 2000.

Longo G., "*On the proofs of some formally unprovable propositions and Prototype Proofs in Type Theory*" Invited Lecture, **Types for Proofs and Programs**, Durham, UK, Dec. 2000; also published in: **Lecture Notes in Computer Science**, vol. 2277 (Callaghan *et al.* eds.), pp. 160–180, Springer, Berlin, 2002.

Longo G., "*The Constructed Objectivity of Mathematics and the Cognitive Subject*", in: **Quantum Mechanics, Mathematics, Cognition and Action** (M. Mugur-Schächter and A. van der Merwe, eds.) Kluwer

Academic Publishers, Dordecht, 2002a.

Longo G., *"The reasonable effectiveness of Mathematics and its Cognitive roots"*, in **New Interactions of Mathematics with Natural Sciences: Geometries of Nature** (L. Boi, ed.), World Scientific, Singapore, 2005.

Longo G., *"Laplace, Turing and the "imitation game" impossible geometry: randomness, determinism and programs in Turing's test,"* in: **Parsing the Turing Test** (Epstein, R., Roberts, G. and Beber, G., eds.), pp. 377–413, Springer, 2007.

Longo G., *"From exact sciences to life phenomena: following Schroedinger and Turing on programs, life and causality,"* special issue of: **Information and Computation**, vol. 207, no. 5, pp. 543–670, 2009.

Longo G., *"Interfaces of Incompleteness"* (downloadable). Preliminary version in French in: **Les mathématiques**, CNRS Editions, 2011.

Longo G. and Tendero P.-E. *"The differential method and the causal incompleteness of Programming Theory in Molecular Biology,"* **Foundations of Science**, no. 12, pp. 337–366, 2007.

Mancosu P. and Ryckman T., *"Mathematics and Phenomenology. The correspondence between Oskar Becker and Hermann Weyl,"* **Philosophia Mathematica** vol. 10, pp. 130–202, 2002.

Mandelbrot B., **The fractal geometry of nature**, W.H. Freeman, New York, 1982.

Merleau-Ponty M., **Phénomenologie de la perception**, Gallimard, Paris, 1945.

Mugur-Schächter M., *"Quantum mechanics versus a method of relativized conceptualization"*, in: **Quantum Mechanics, Mathematics, Cognition and Action**, (M. Mugur-Schächter and A. Van der Merwe, eds.) Kluwer Academic Publishers, Dordecht, 2002.

Mugur-Schachter M., **Sur le tissage des connaissances**, Hermès-Lavoisier, Paris, 2006.

Nelson E., *"Internal set theory,"* **Bulletin of the American Mathematical Society**, vol. 83, no. 6, p. 1165, 1977.

Nicolis G., *"Dissipative systems"*, **Reports on Progress in Physics**, vol. 49, no.8, p. 873, 1986.

Nicolis G. and Prigogine I., **Self-organization in non-equilibrium systems**, Wiley, New York, 1977.

Nicolis G. and Prigogine I., **A la rencontre du complexe**, Presses Universitaires de France, Paris, 1989.

Noether E., *"Invariante Variationsprobleme,"* **Nachrichten von der**

Königlicher Gesellschaft den Wissenschaft zu Göttingen, **Math-phys. Klasse**, no. 2, pp. 235–257, 1918.

Nottale L., **La relativité dans tous ses états**, Hachette, Paris, 1999.

Novello M., **Le cercle du temps**, Atlantisciences, Paris, 2001.

Pachoud B., "*The Teleological Dimension of Perceptual and Motor Intentionality*", in: **Naturalizing Phenomenology** (Petitot et al., eds.), Stanford University Press, 1999.

Parrini P., **Conoscenza e Realta**, Laterza, Roma-Bari, 1995.

Patras F., **La pensée mathématique contemporaine**, Presses Universitaires de France, Paris, 2001.

Pauri M., **I rivelatori del tempo**, preprint, Diploma di Fisica, University Di Parma, 1999.

Peters R., **The ecological implication of body size**, Cambridge University Press, 1983.

Petit J.-L., "*La spatialitÃl' originaire du corps propre: phenomenology et neurosciences,*" **Numéro spécial de Revue de Synthèse**, t. 124 (Longo, ed.), Editions de la rue d'Ulm, Paris, 2004.

Petitot J., "*Idéalités mathématiques et réalité objective. Approche transcendantale,*" in: **Hommage à Jean-Toussaint Desanti** (G. Granel, ed.), Editions TER, Mauvezin, 1991.

Petitot J., "*Préface*" in **Ecrits Philosophiques**, G. Preti (traduction et présentation de Luca Scarantino), Editions du Cerf, Paris, 2002.

Petitot J., **Neurogéométrie de la vision**, Les Editions de l'Ecole Polytehcnique, Paris, 2008.

Petitot J., Varela F., Pachoud B. and Roy J.-M. (eds.), **Naturalizing Phenomenology: issues in comtemporary Phenomenology and Cognitive Sciences**, Stanford University Press, 1999.

Pilyugin S.Y., **Shadowing in dynamical systems**, Springer-Verlag, Berlin, 1999.

Poincaré H., **La Science et l'Hypothèse**, Flammarion, Paris, 1902.

Preti G., **Saggi filosofici, vol. I**, La Nuova Italia, Firenze, 1976.

Prigogine I. and Stengers I., **Entre le temps et l'éternité**, Fayard, Paris, 1988.

Prochiantz A., **Les anatomies de la pensee**, Odile Jacob, Paris, 1997.

Reinberg A., **Les rythmes biologiques**, Presses Universitaires de France, Paris, 1989.

Ricard J., "*Emergence, organisation et causalité dans les systèmes biologiques,*" in: **Enquête sur le concept de causalité** (L. Viennot and C. Debru, eds.), Presses Universitaires de France, Paris, 2003.

Riemann B., **On the Hypotheses which Lie at the Bases of Geometry**, translated by W.K. Clifford, Nature, vol. 8, pp. 14–17, 1873.
Robinson A., **Non standard analysis**, North-Holland, Amsterdam, 1980.
Rogers H., **Theory of Recursive Functions and Effective Computability**, MacGraw Hill, New York, 1967.
Rosen, R., **Life Itself**, Columbia University Press, New York, 1991.
Rosenthal V. and Visetti Y.-M., "*Sens et temps de la Gestalt*", **Intellectica**, vol. 28, pp. 147–229, 1999.
Ryder L.H., **Quantum field theory**, Cambridge University Press, 1986.
Sakharov A. D., **Oeuvres scientifiques**, Anthropos, Paris, 1982.
Salanskis J-M., **L'herméneutique formelle**, CNRS Editions, 1991.
Sauer T. "*Shadowing breakdown and large errors in dynamical simulations of physical systems*", Phys. Rev. E65, 036220 (2002).
Schmidt-Nielsen K., **Scaling**, Cambridge University Press, 1984.
Seldin J.P., "*Curry's program*" in **To H.B. Curry: essays on combinatory logic, lambda-calculus and formalism**, (J.P. Seldin and J. R. Hindley, eds.), Academic Press, London, 1980.
Sherrington D., "*Landscape paradigms in physics and biology: introduction and overview,*" **Physica D**, vol. 107, p. 117, 1997.
Simpson G.G., **The Major Features of Evolution**, Harvard University Press, Cambridge, MA, 1953.
Solé R. and Goodwin B., **Signs of Life**, Basic Books, New York, 2000.
Soto A. and Sonnenschein C., "*The somatic mutation theory of cancer: growing problems with this paradigm?*" **BioEssays**, vol. 26, p. 1097–1107, 2004.
Stewart J., "*La modélisation en biologie,*" in: **Enquête sur le concept de modèle** (P. Nouvel, ed.) Presses Universitaires de France, Paris, 2002.
Tappenden J., "*Geometry and generality in Frege's philosophy of Arithmetic*" Synthese, vol. 102, no. 3, 1995.
Tazzioli R., "*Riemann: alla ricerca della geometria della natura*", **Le Scienze**, no. 14, supplement 380, 2000.
Teissier B., "*Protomathematics, perception and meaning of mathematical objects*" in: **Images and Reasoning**, (Okada et al., eds.), Keio University Press, Tokyo, 2005.
Thom R., **Stabilité structurelle et Morphogénèse**, Benjamin, Paris, 1972.
Thom R., **Modèles mathématiques de la morphogenèse**, Christian Bourgois, Paris, 1980.
Thompson W.D., **On growth and form**, Cambridge University Press,

1961.

Tieszen R., **Phenomenology, Logic, and the Philosophy of Mathematics**, Cambridge University Press, 2005.

Tonietti T. L., "*Four letters of E. Husserl to H. Weyl and their context*" in: **Exact Sciences and their philosophical foundations**, Peter Lang, Frankfurt, 1988

Troelstra A.S., **Metamathematical investigation of Intuistionistic Logic and Mathematics**, LNM 344, Springer-Verlag, Berlin, 1973.

Turing A., "*Computing Machines and Intelligence*", **Mind**, vol. LIX, 1950.

Ullmo J., **La pensée scientifique moderne**, Flammarion, Paris, 1969

Varela F., **Autonomie et connaissance**, Seuil, Paris, 1989.

Varela F., "*The specious Present: A Neurophenomenology of Time Consciousness*" in: **Naturalizing Phenomenology** (Petitot et al. eds.), Stanford University Press, 1999

Verlet L., **La malle de Newton**, Gallimard, Paris, 1993

Victorri B., "*Espace sémantiques et représentation du sens,*" **Textualités et nouvelles technologies, éc/artS**, vol. 3, 2002.

Vidal C. and Lemarchand H., **La réaction créatrice. Dynamique des systèmes chimiques**, Hermann, Paris, 1988.

Viennot L., "*Raisonnement commun en physique: relations fonctionnelles, chronologie et causalité*" in: **Enquête sur le concept de causalité** (L. Viennot and C. Debru, eds.) Presses Universitaires de France, Paris, 2003.

Vogel G. and Angermann H., **Atlas de la biologie**, Le Livre de Poche, LGF, 1994.

Waddington C. H., "*Stabilization of Systems, Chreods and Epigenetic Landscapes*", **Futures**, vol. 9 no. 2, p. 139, 1977.

Wang H., **Reflections on Kurt Gödel**, MIT Press, Cambridge, MA, 1987.

West G.B., Brown J.H. and Enquist B.J., "*A general model for the origin of allometric scaling laws in biology*", **Science**, vol. 276, p. 122, 1997.

Weyl H., **Das Kontinuum, Kritische Untersuchungen uber die Grundlagen. der Analysis**, Veit, Leipzig, 1918a. (published in English as: **The Continuum: A Critical Examination of the Foundation of Analysis**, translated by S. Pollard and T. Bowl, Dover, New York, 1994).

Weyl H., **Raum–Zeit–Materie**, Springer, Berlin, 1918b.

Weyl H., **Philosophie der Mathematik und Naturwissenschaft**, München, Verlag, Leibniz, 1927. (published in English as: **Philoso-

phy of Mathematics and Natural Science, Princeton University Press, 1949).

Weyl H., **Symmetry**, Princeton University Press, 1952.

Young *et al.*, **Nature**, 400, pp. 766–788, 1999.

Zurek W.H., "*Decoherence and the Transition from the Quantum to the Classical*", **Physics Today**, vol. 44, p. 36–44, 1991.

Index

abstraction, mathematical, 63
allometry, 167
ameba, 61
analogy, 166
anti-foundationalism, 283
approximation
   digital, 206, 207
   geometric, 208
   of measurement, 277
arrow of time, 189
Aspect, 87
asymmetrical reading of the formal determination, 191, 192
asynchronous processing, 114
Atlan, 127
Automatic Theorem Proving, 78
autopoiesis, 6, 244
axiom of choice, 2
axiomatic system, 69
axiomatic, sense of, 283

Badiou, 28
Bak's sand pile, 232
Bell inequalities, 87
Bernard, 132
biolon, 257
   stability of, 255
Bitbol, 295
Boltzmann constant, 273
Brouwer, 41, 72
Brouwer's solipsism, 41

calculation, the concept of, 14
Cantor, 2
capacity to forget, 42, 67
Cartesian "Ego", 38
Cartesian cogito, 82
Cassirer, 24
Category Theory, 34, 40
causal law, 53
causality, 54, 55, 140, 161, 185
   epistemic, 188
   local/global, 246
   material, 189
   objective, 188
   physical, 182
cause
   efficient, 185, 190, 191
   material, 185
chaos, deterministic, 262, 264, 286
Châtelet, 58, 76, 82
Chevalley, 194
chirality, 132
Church, 128
circularity, 35
classical ascription of probabilities, 275
classical statistics
   Maxwell–Boltzmann, 272
cognition as sequence-matching, 59
cognitive foundations, 20
cognitive relativism, 24
coherent structure, 232, 233
concept, scientific, 46

conceptual renormalization, 145
Connes, 93, 109
construction
    of objectivity, 162
    the concept of, 12
contingent finality, 137, 212
continuity, topological, 96
continuous line, 68
continuum, 1, 80
continuum hypothesis, 15
correlation length, 62, 242
CPT theorem, 157
critical phenomena, 228
critical system, 142
critical theories, 193
critical transition, 229
criticality, 97
cybernetics, 125

Darwinian evolutionary theory, 121
decidability, 85
definition of mathematics, 64
degeneracy, 213
    functional, 213
    systemic, 213
determination
    formal, 184
    objective, 184
determination vs cause, 183
determinations involved in living
    phenomena, 169
determinism, 123
digital simulation vs mathematical
    modeling, 202
dimension as a topological invariant,
    129
dimensional analysis, 129
discrete, definition of, 259
discrete logistic equation, 207
dualism, 173
dynamic system, 87
dynamical system, 121
    non-linear, 110
    theories of, 112

emergence of complex structures, 231

energy, 133
entropy, 249
    negative, 249
EPR argument, 87, 90, 108
equivalence of different formalizations
    of computability, 128
equivalence relation, 159, 160
Euclid, 81, 93
Euclid's axiomatics, 43
evolution of a calculus in physical
    terms, 205
evolutionary explanation, 122
extended criticality, 120, 133, 235,
    253, 293

Fermat's last theorem, 77
Fibonacci sequence, 126
finalism, 248
    in physics, 250
finite/infinite, 75, 95
flux
    of energy, 217
    of information, 217
force, 54
formal, 65
formalism, 2, 4, 7, 32, 98
    formal incompleteness of, 73
foundation, 8, 36, 57
foundational analysis, aim of, 284
foundations, 21
    of mathematics, 5, 17, 42, 56, 69,
        81, 197
    of physics, 5
Frege, 2, 21, 70
friction with the world, 45
functionalism, 105
functionalist account of cognition, 104
functor, 34

Galileo, 150
gauge theories, 54
Gauss, 69, 77
Gauss's proof, 78
gene, 127
genesis, 21
geodesic, 137, 234

geodesic principle, 149, 151, 154, 156
geometric judgment of well order, 75, 76
geometric vs algebraic approach, 113
geometrization, 50, 52
geometry
  as a human construction, 93
  Euclidean, 156
  for Husserl, 26
  fractal g. of organs, 134
  non-commutative, 93, 112
geometry of proof, 66, 84
gesture, 61, 66, 69, 80, 84
  mathematical, 58
Girard, 41, 66, 84
Girard's proof analysis, 41
Gödel, 26, 27, 32, 65
Gödel incompleteness theorem, 287
Gould, 137

Heisenberg algebra, 109
Heyting, 80
hidden variables in quantum mechanics, 90
Hilbert, 2, 70, 85
homology, 166
homotheties, 22
Husserl, 22, 23, 34, 82

identification between algorithm and law, 204
imitation, computational, 203
incompleteness, 85, 91
  Gödel's theorem of, 33
  of quantum physics, 88
  of the formal theory of numbers, 29
  theorem, 73
indetermination, 91, 288
  biological, 251, 289
  quantum, 91
inertia, 150
infinite ML-random sequences, 278
infinity, actual, 94, 96
information, 53, 128, 218
  as deformation of geometric structure, 130

integers, 70, 71
integer, concept of, 76
  sequence of, 70
  well-order of, 71
integration, 247
intentionality, 119, 130
internal clocks, 136
internal/external, 145
  articulation of space, 138
intuition, 79, 80
intuitionism, 41
invariance, 9, 18, 51, 194
  gauge, 54
invariant, 4, 22, 29, 34, 42, 67, 68, 120, 138, 153
  biological, 219, 253
  gauge, 38
irreversibility
  of thermodynamic, 95
  of time, 111, 113
iterative vs processual, 141

Kaluza–Klein theory, 107
Kant, 144
Kastler, 289
Kleene, 128
knowledge, exact, 20

lambda-calculus, 66
landscape paradigm, 156
language, natural, 51
Laplace, 85, 103
Laplace's hypothesis, 198
Laskar, 263
law
  fundamental l. of physics, 151
  logical, 282
  notion of, 204
laws of conservation, 187
laws of thought, 21, 69
local vs global, 139
logicism, 2, 4
logicist dogma, 21
Lowenheim, 70

mathematical constitution of

phenomena, 107
mathematical models of biological systems, 132
mathematical object, assumption of, 26
mathematical objectivity, construction of, 18
mathematical structures, 33
mathematization, 32, 49
mean field theories, 135
mechanica universalis, 104
monism, 173, 238
morphism, 34
morphogenesis, 125, 131
   Turing's model of, 203
morphogenetic field, 125

negative result, 286
Nernst principle, 269
Newton, 49, 226
Noether's theorem, 193
non-commutativity, 91
non-local effects, 108
non-locality, 267, 268
   as globality in classical system, 271
non-separability, 267, 268
non-standard analysis, 14
number line, 71, 76
number theory, 79
number, concept of, 72

objective, 47
optimality, 233
order parameter, 98
order, spontaneous formation of, 234
organizational enclosure, 237
orgon, 257

passing from local to global, 229, 230
Pasteur, 132
pathology, causes of, 214
Peano, 70
phase transition, 94, 98, 227
   theories of, 110
physical singularity of life, 235
physical vs living phenomena, 173

Piero della Francesca, 76
Planck constant, reduction to 0, 271
Platonism, 65
Poincaré, 66
Poincaré–Berthoz isomorphism, 67
positivism, 5
power laws, 167
   examples of, 167
predictability, 198
principle
   Curie, 55
   logical, 2
   of construction, 2, 4, 6, 31, 32, 73, 89
   of least action, 149
   of optimization, 168
   of proof, 2–5, 31, 73, 89
   of the shortest path, 149
   teleonomic, 6
proof, 84
property, 23, 25, 184

quantum mechanics, 107
quantum non-separability, 91
quantum statistics
   Bose–Einstein, 272
   Fermi–Dirac, 272
quantum theory of the hydrogen atom, 96
quantum vs classical statistics, 272

randomness
   biological, 254
   classical, 262, 269
   epistemic, 264
   epistemic notion of, 201
   in algorithmic theory of information, 201
   Martin-Löf mathematical definition, 278
   ML-r., 279
   objective, 265
   objective quantum, 267
   quantum r. as epistemic, 270
rationality, 59
realism, 40

in mathematics, 28
reality, 45
reduction of biology to physics, 226
reference frame, 38, 40
referring function of language, 48, 50
regular vs singular, 142
regulation, 247
relativity, 54, 105
   general, 53, 106
   special, 105
relaxation processes, 136
renormalization, 95, 228
resistance to changes in viewpoint
   due to accountability obligation, 290
   philosophical, 289
Riemann, 34, 69, 93
rigor of proof, 63
rule, formal, 2

saccade, 67, 119
scale-invariant parameter, 133
scaling, 167
   rules of, 219
self-organization, 231
   critical, 232
semantics, 89
sensitivity to the initial conditions, 199
sequence matching, 60
Set Theory, 40, 81
signification, constitution of, 61
singularity, 142
Skolem, 70
SNARC effect, 71
space, 163
   as group, 143
   as the form of external sense, 144
spatial dimensions for biology, 171
state, 184
subject, cognitive, 39
subjective, 47
super-symmetric theories, 109
symbol, 63
symmetry, 4, 52, 153, 160, 248
   external, 192

   internal, 193
symmetry breaking, 33, 52, 98, 155, 193
   in critical phenomena, 155
   spontaneous, 54, 154
syntax, 89

Teissier, 67
temporality, 35, 138, 142
   anticipatory form of, 137
   biological, 136
   internal, 136
   spatialized type of, 141
theorem, 43
   proving of, 77
theoretical characterization of physical objects, 184
thermodynamics, 110
Thom, 126
three-body problem, 85, 110, 113
three-dimensionality of living systems, 134
time, 163
   as semi-group, 143
   as the form of internal sense, 144
   bi-dimensionality of, 243
   genetic-structural, 117
   of clock mechanisms, 115
   of constitutive process, 116
   phenomenal, 115, 118
   relational, 117
topological evolution, 126
transcendence, 23
transcendental constitution of mathematical objectivities, 97
transportation process
   diffusive, 242
   propagative, 242
truth
   ontological, 66
   transcendental, 39
Turing, 104, 128, 203
Turing machine, 104

uncertainty principle, 91
undecidability, 86, 280

undecidability theorem, 73
unification, 226, 288
  of biology with physics, 122
unpredictability, 280
  of a deterministic system, 264
  of chaotic dynamic systems, 86
  of Poincaré, 86
  of quantum theories, 87

van Fraassen, 53, 160
vision as palpation through looking, 81

wave–particle duality, 25
Weyl, 34, 37, 39, 51, 116
Wittgenstein, 22
writing, origins of, 83